GERMANY-BASED ENCRYPTED MESSAGING APP TELEGRAM EMERGES AS JIHADIS' PREFERRED COMMUNICATIONS PLATFORM:

Part V of MEMRI Series:
Encryption Technology Embraced by ISIS, Al-Qaeda, Other Jihadis

(September 2015–September 2016)

Steven Stalinsky

ميمري
**MEMRI
BOOKS**

MEMRI Books
Washington, D.C.

Germany-Based Encrypted Messaging App Telegram Emerges as Jihadis' Preferred Communications Platform:

Part V of Memri Series: Encryption Technology Embraced by ISIS, Al-Qaeda, Other Jihadis (September 2015-September 2016)

Published in the United States of America by MEMRI Books

www.memri.org | cjlab.memri.org

ISBN 978-0-9678480-6-8 (paperback)
ISBN 978-1-7344283-0-8 (e-book)
Library of Congress Control Number: 2020900825

Table of Contents

NEW MAJOR MEMRI REPORT ON TELEGRAM IS BASIS OF 'WASHINGTON POST' DECEMBER 24 FRONT-PAGE STORY, WHICH QUOTES REPORT IN HEADLINE – "THE 'APP OF CHOICE' FOR JIHADISTS" – EXTENSIVELY QUOTES MEMRI EXECUTIVE DIRECTOR, AND INCLUDES PHOTO OF MEMRI JTTM RESEARCHERS

The Washington Post

THE 'APP OF CHOICE' FOR JIHADISTS: ISIS SEIZES ON INTERNET TOOL TO PROMOTE TERROR

BY: JOBY WARRICK; December 23, 2016

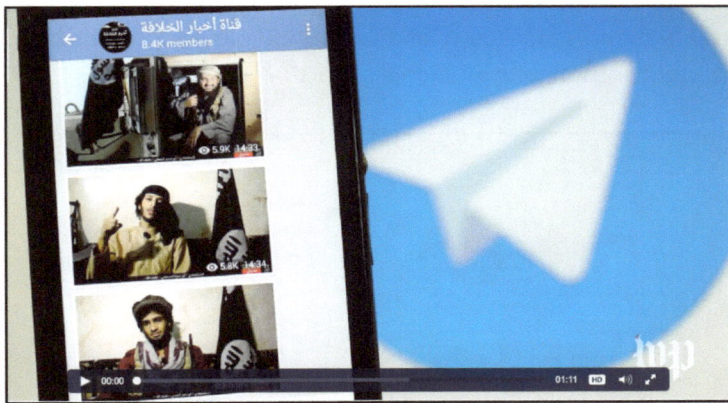

Telegram, the high-security messaging app, has shut down 78 Islamic State-related channels. Here's a look at what the app is and why terrorists are using it. (Jhaan Elker/The Washington Post)

When the Islamic State was seeking volunteers for a holiday killing rampage in Europe, it sent word over its favorite social-media channel: the messaging service known as Telegram.

"Christmas, Hanukkah, and New Years Day is very soon," began a Dec. 6 posting on one of the terrorist group's usual Telegram bulletin boards. "So let's prepare a gift for the filthy pigs/apes."

Two weeks later, when a truck mowed down pedestrians at a crowded Berlin Christmas market, the group again used Telegram, this time to claim credit for the attack. On Friday, after chief suspect Anis Amri was killed in a Milan shootout, Telegram broadcast his posthumous video. The Tunisian migrant had fled Berlin and crisscrossed France and Italy before being stopped by Italian police looking for a burglary suspect. In his video he pledges allegiance to the Islamic State and issues a chilling warning to Westerners: "God willing, we will slaughter you."

The words and images flew across the globe over a network that terrorist leaders describe as ideal for their purposes — one that is highly discreet, with its heavy encryption and secret chat rooms, but also highly permissive, allowing violent Islamist groups to exchange ideas and spread propaganda with minimal interference. The same conclusion has been reached by terrorism analysts who say Telegram is now overwhelmingly preferred by extremist groups such as the Islamic State, in part because the company has failed to adopt the aggressive measures used by its competitors to kick terrorists off its channels.

A report this week by an organization that monitors Islamist militants' Internet communications calls Telegram "the app of choice for many ISIS, pro-ISIS and other jihadi and terrorist elements." The study describes the terrorists' mass migration to Telegram as one of the most striking developments in the field recently. ISIS is one of the common acronyms for the Islamic State.

"It has surpassed Twitter as the most important platform," said Steven Stalinsky, lead author of the report by the Washington-based Middle East Media Research Institute, also known as MEMRI. "All the big groups are on it. We see ISIS talking about the benefits of Telegram and encouraging its followers to use it."

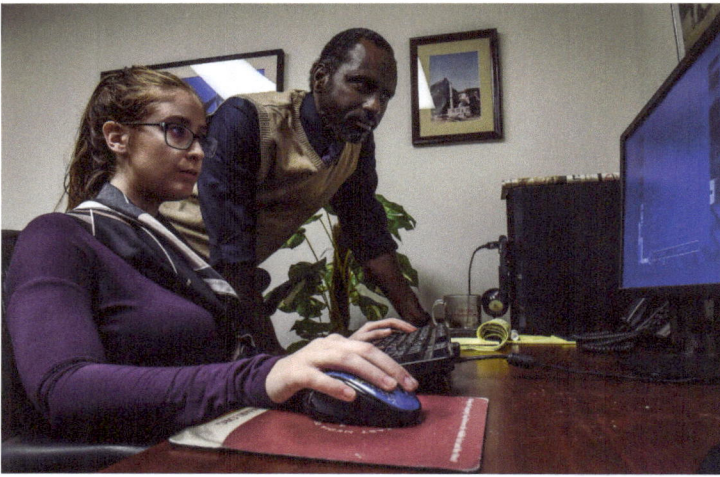
Mansour Al-Hadj, top, and Anat Agron, part of MEMRI's Jihad and Terrorism Threat Monitor project, monitor the social-media site Telegram on Dec. 22 in their office in Washington. (Bill O'Leary/The Washington Post)

Terrorists' use of Telegram has been a growing concern among U.S. and European counterterrorism officials for more than a year, as well as a source of numerous inquiries and complaints lodged against the German-based company and its creator, Pavel Durov, a 32-year-old Russian national who launched the service in 2013 with his brother Nikolai.

Just three days before the assault on the Berlin Christmas market, senior members of the House Foreign Affairs Committee urged Durov to immediately take steps to block content from the Islamic State, warning that terrorists were using the platform not only to spread propaganda but also to coordinate actual attacks.

"No private company should allow its services to be used to promote terrorism and plan out attacks that spill innocent blood," stated the letter, signed by Rep. Ted Poe (R-Tex.), the chairman of the panel's subcommittee on terrorism and nonproliferation, and Rep. Brad Sherman (Calif.), the ranking Democrat on the subcommittee on Asia and the Pacific.

Efforts this week to reach Telegram's founder through his social-media accounts were unsuccessful.

Durov, who fled Russia in 2014 and now lives in a kind of self-imposed exile as a citizen of the island state of Saint Kitts and Nevis, has in past interviews and essays defended his company's efforts at self-policing, noting that Telegram shut down 78 channels used by the Islamic State in the wake of the Nov. 13, 2015, terrorist attack in Paris. In March, Durov told CBS's "60 Minutes" that he was "horrified" to see terrorist groups infesting

Telegram's chat rooms, and he said the company was trying to do more to stop them.

But Durov also contends that it is impossible to fully prevent terrorists from taking advantage of the encrypted communication services Telegram offers to its 100 million active users, a global network that includes millions of people living in countries that deny citizens the right to free expression.

"There's little you can do, because if you allow this tool to be used for good, there will always be some people who would misuse it," Durov told CBS.

But critics of the company — a cohort that includes high-ranking U.S. counterterrorism officials — say Telegram could do more. A primary reason for the German firm's popularity with violent groups is the fact that rival social-media companies have aggressively cracked down on them, U.S. officials say. Facebook and Twitter — two firms that were once criticized for allowing terrorist postings on their pages — have received high marks in recent months for their efforts to find and block radical Islamist content as soon as the material surfaces.

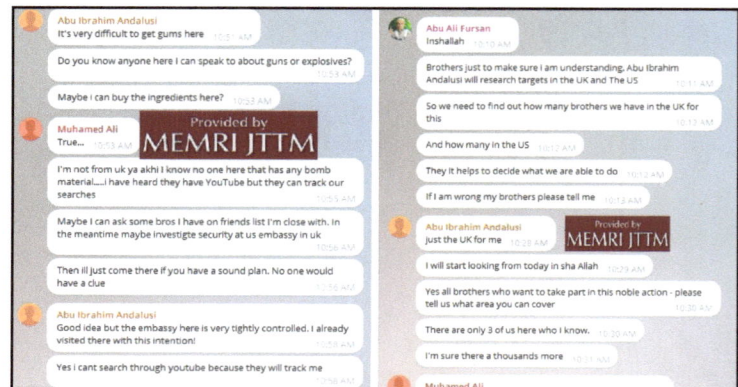

"Positive steps by Twitter, for example, are part of the reason Telegram is becoming the new thing," said a senior administration official involved in tracking the Islamic State's online presence. The official, who spoke on the condition of anonymity to discuss sensitive analysis of the terrorist group's operations, called the migration to Telegram a "major cause of concern," in part because of encryption features that make it harder for law enforcement officials to discover and thwart the terrorists' plans.

"It's alarming because it shows they're really good at adapting to new means," the official said. To stop attacks, private companies and government officials must stay a step ahead of the terrorists and "figure out how to deny them these capabilities before they even start using them," he said, adding, "That simply hasn't been the case with Telegram."

New easy-to-use features installed by Telegram have made the task of preempting terrorism even harder, government officials and private experts say.

Originally a phone-based software with a relatively small but devoted following, Telegram last year introduced a new version for desktop computers that made it easier to transmit videos and large files as well as private messages.

New "end to end" encryption was added in April to give users an extra assurance of privacy. Independent analysts have described the quality of Telegram's encryption as "military grade," meaning that it is extremely difficult, if not impossible, to crack. Users can also opt for a self-destruct feature in which private messages disappear as soon as they are read.

MEMRI's Stalinsky monitors hundreds of jihadist-related Telegram channels from a bank of computers at his office, watching live "chats" joined by individuals who sign on to forums linked to the Islamic State, al-Qaeda and dozens of other groups. Often, he says, a participant in one of the open discussions will signal that he wants to have a private conversation. That typically means joining a temporary, invitation-only "secret" chat group that will exist for only a few hours and then disappear.

"If you're not watching at that precise moment, you'd never know about it," Stalinsky said.

Even the more public conversations often convey specific instructions about bombmaking or potential targets for terrorist attacks. In recent months, Islamic State leaders and supporters have posted messages on Telegram containing lengthy "kill lists" of Westerners the terrorists sought to mark for execution, as well as appeals to sympathetic scientists and engineers to join the Islamic State's efforts to produce advanced weapons.

Last month, Stalinsky, using an anonymous user name, gained entrance to a secret Telegram chat in which self-described British and American supporters of the Islamic State discussed ideas for attacking the U.S. Embassy in London.

Stalinsky alerted U.S. officials to the conversation, and there is no known evidence suggesting that the would-be terrorists put their plan into motion. But images of the text exchange show the participants discussing in detail the logistics for such an attack, including weapons and travel arrangements. The individuals even discuss whether they should take steps to avoid killing women and children.

"Women and children should be off limits," one of the participants wrote. "We do not want to be like kuffar [infidels]."

The fact that the group allowed a complete stranger to monitor the chat suggests that the plotters were not true Islamic State operatives. While Telegram's users include senior terrorist leaders and operatives, many chat-room inhabitants appear to be merely fanboys and wannabes, and some clearly "are not that smart," Stalinsky said.

Yet, even bumblers are capable of striking a blow for the Islamic State. And using Telegram, Stalinsky said, the naive and willing have opportunities to connect with professionals — experts at the tools of terrorism and the use of social media — to put deadly plans into action.

"The West has been generally two steps behind the jihadis when it comes to cyber," Stalinsky said. "Many people in government are still focused on Twitter, and they need to be. But what we tell them is, 'That's no longer the main story.'"

Stills from video that accompanied the Washington Post *article's online version*

DAYS AFTER MEMRI EXPOSES TELEGRAM AS JIHADIS' 'APP OF CHOICE' – IN 'WASHINGTON POST' ON DECEMBER 24 – TELEGRAM CREATES ISIS WATCH CHANNEL ALLEGING IT IS REMOVING ISIS CONTENT

On December 26, 2016, days after MEMRI exposed Telegram as ISIS's and other jihadis' "app of choice" – as published by the *The Washington Post* on its front page on December 24 – Telegram responded by creating its ISIS Watch channel alleging, in daily updates, that the platform is removing ISIS content. As of January 3, the channel has 3,357 members.

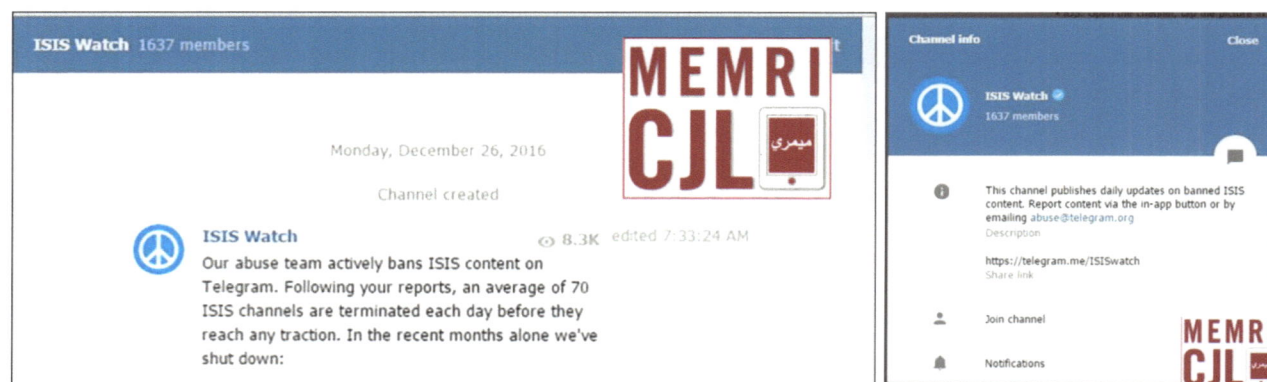

Channel information for ISIS Watch

ISIS Watch Channel: "Our Abuse Team Actively Bans ISIS Content"

The channel states: "Our abuse team actively bans ISIS content on Telegram. Following your reports, an average of 70 ISIS channels are terminated each day before they reach any traction. In the recent months alonw we've shut down: September: 1987 ISIS bots and channels; October: 1800 ISIS bots and channels; November: 2094 ISIS bots and channels; December 2013 ISIS bots and channels... " It also gives instructions about how to report content to the Telegram abuse team.

Both Pavel Durov's Twitter account and the official Telegram account announced the new channel. The following are related Telegram and Twitter announcements and posts.

Within a day of its launch, the ISIS Watch channel had over 1,600 members; a few days later, that number had more than doubled.

Telegram Seems To Be Doing What Twitter Did For Years – Taking Little Or No Real Action Against Terrorist Content

ISIS It should also be noted that ISIS Watch provides no proof that Telegram is actually removing or banning the content that it claims to be removing or banning. The company seems to be doing what Twitter did before it with regard to removing terrorist content: Over the course of a couple of years, Twitter claimed numerous times that it was doing this but was taking little or no real action, and allowed it to continue to proliferate until the horrific video of the beheading of James Foley and other American and Western journalists was disseminated via the platform in August 2014. Following this, Twitter changed its approach and began to severely curb terrorist content on its platform, to the point where many jihadis have tired of having accounts shut down and will not even try to use it any more. In contrast, Telegram's first announcement that it had removed any terrorist content at all – a mere 78 ISIS-connected public channels – came only after the widescale Paris attacks in November 2015 whose planners were found to have used the platform to coordinate the attacks. It is going to take government pressure, and perhaps also media pressure, to push Telegram to take real and verifiable action to remove terrorist content on its platform.

ISIS Watch Posts: Information, Announcements, Reports; Posts And Claims By Durov

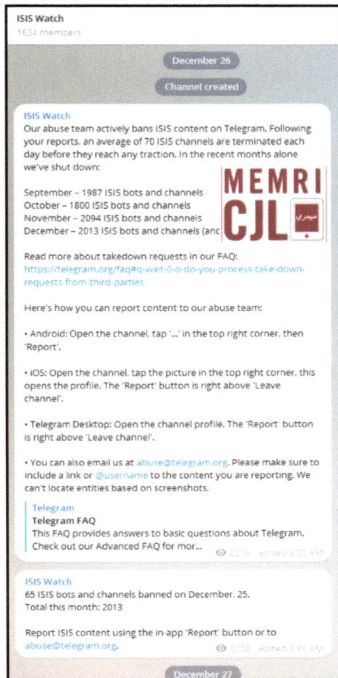

Announcement of channel's creation and preliminary report on bots and channels that have been banned

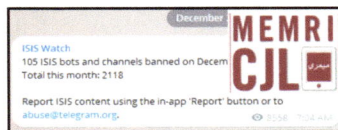

Report on content banned on December 26, 2016

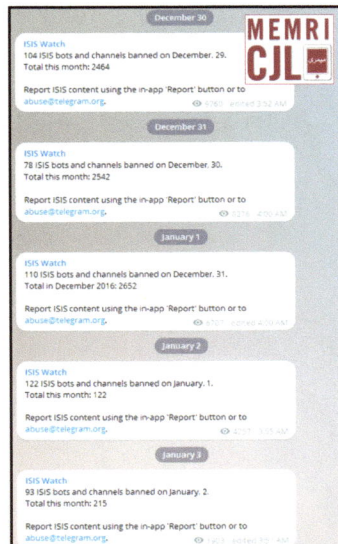

Reports of content banned December 30, 2016-January 3, 2017.

Telegram's Twitter announcement of ISIS Watch

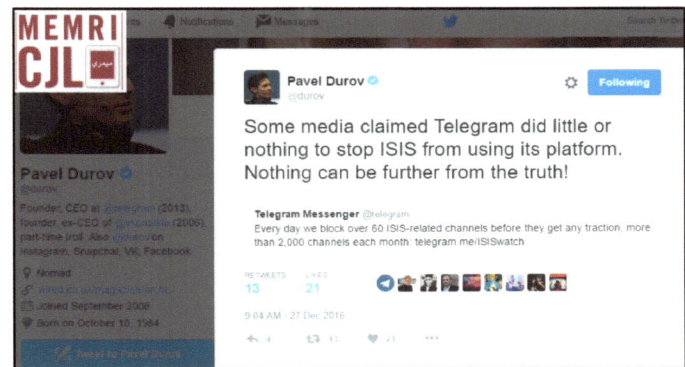

Durov states that Telegram is taking steps to stop ISIS on the platform, and retweets Telegram Twitter announcement

Durov responds to a tweet to him of Washington Post story from journalist Ariel Ben Solomon

IN CONGRESSIONAL LETTER TO DUROV THAT INCLUDES RESEARCH EXCLUSIVELY PROVIDED BY MEMRI, HOUSE TERRORISM SUBCOMMITTEE CHAIRMAN AND RANKING MEMBER UNDERLINE "GRAVE CONCERNS ABOUT... FOREIGN TERRORIST ORGANIZATIONS" ON PLATFORM, EXHORT HIM TO "DO ALL IN YOUR POWER TO PREVENT TERRORISTS" FROM USING IT TO "ADVANCE THEIR LETHAL CAUSE"

Congress of the United States
Washington, DC 20515

December 16, 2016

Pavel Durov
Founder and CEO
Telegram

Dear Mr. Durov:

We write to communicate our grave concerns about reports regarding the recent migration of Foreign Terrorist Organizations (FTOs) and their supporters from other social media platforms to Telegram. Terrorists are reportedly using Telegram's encrypted service to disseminate propaganda, drive fundraising, recruit new members, and coordinate attacks.

As U.S. Members of Congress, we are strong advocates for the right to privacy and appreciate Telegram's strong commitment to protecting this right. However, no private company should allow its services to be used to promote terrorism and plan out attacks that spill innocent blood. So we were thankful to see your removal of 78 ISIS-related channels across 12 languages on the heels of last year's attacks in Paris. This, however, is just a start.

Hundreds of channels affiliated with ISIS and other terrorist organizations still find refuge in Telegram's encrypted service. "Nashir News Channel", an outlet of ISIS that delivers pro-ISIS news in several languages and has been suspended on Twitter and YouTube, attracts well over 10,000 regular followers on Telegram. Other ISIS channels have issued calls to target U.S. military bases in the Middle East and facilitated the selling of two prisoners, one from Norway and another from China. "Kill lists" distributed this past summer by ISIS on Telegram included names, addresses, and other personal details of hundreds of U.S government personnel, police officers, and employees of major U.S. companies. Al-Qaeda in the Arabian Peninsula (AQAP), one of the most lethal terrorist groups in the world, uses Telegram to release its statements and propaganda. Other terrorist groups such as Hamas, Jabhat Al-Nusra, and the Taliban are also on Telegram. We are troubled at these examples and many others that illustrate how your platform is allowing the spread of terrorist ideology and the coordination of actual terrorist attacks.

Telegram is not the first to face this challenge. It is our hope that Telegram will learn from other social media companies who have successfully maintained their emphasis on privacy while reducing the terrorist footprint on their platforms. They did this by making significant investments over a sustained period of time. These investments included making it easier for users to report terrorist content, having dedicated teams to examine reports and proactively search for terrorist content, using algorithms and other automated mechanisms to assist manual reviews, quickly identifying and removing accounts promoting terrorism, regularly training review teams on what to look for and new ways terrorists are using the platform, and, most recently, coming together to form a shared database of terrorist content.

Terrorists will continue to try to kill innocent people even if social media companies never existed, but it is in the interest of us all to not enable and even support their nefarious actions. We respectfully request that you do all in your power to prevent terrorists from exploiting Telegram to advance its lethal cause.

Sincerely,

Ted Poe
Chairman, Subcommittee on Terrorism,
Nonproliferation, and Trade

Brad Sherman
Ranking Member, Subcommittee on Asia
and the Pacific

GERMANY-BASED ENCRYPTED MESSAGING APP TELEGRAM EMERGES AS JIHADIS' PREFERRED COMMUNICATIONS PLATFORM – PART V OF MEMRI SERIES: ENCRYPTION TECHNOLOGY EMBRACED BY ISIS, AL-QAEDA, OTHER JIHADIS – SEPTEMBER 2015-SEPTEMBER 2016

By: Steven Stalinsky and R. Sosnow*

Introduction

The following is Part V in MEMRI's series on jihadis' use of encryption technology.[1] Since the publication of Part IV of this series, on June 16, 2015,[2] the Germany-based encrypted messaging app Telegram has emerged as jihadis' preferred app for encrypted communications. Not since jihadis began joining Twitter en masse has there been such a development.

During the past year, Telegram has been revealed as the channel of communication in numerous terrorist attacks, including the November 2015 Paris attacks, and jihadis planning attacks using it have been arrested worldwide. Telegram's emergence has complicated efforts by the FBI, European security forums, and other agencies to monitor the growing amount of jihadi content online and offers an alternative to platforms such as Twitter, Facebook, and others that are more aggressive in monitoring and removing such content.[3] This situation is another step towards the future scenario described by FBI Director James Comey in October 2014 as law enforcement agencies "going dark" because of expanding encryption capabilities.[4]

A year ago, ISIS, Al-Qaeda, and other jihadi groups and individuals favored Surespot, Kik, and Wickr for encrypted and private communications. WhatsApp was also very popular, even before it offered end-to-end encryption, which came only in April 2016.

Now, Telegram has become the app of choice for many ISIS, pro-ISIS, and other jihadi and terrorist elements; launched in 2013, it was not widely used by jihadis at first. It did not feature prominently in Part IV of the MEMRI series on encryption, and was barely mentioned in Part III, published February 4, 2015;[5] it has now emerged as a major factor in jihadi communications. Often, jihadi groups and individuals on Telegram disseminate the same content on other platforms such as Twitter, and constantly refer and link users to these parallel accounts.

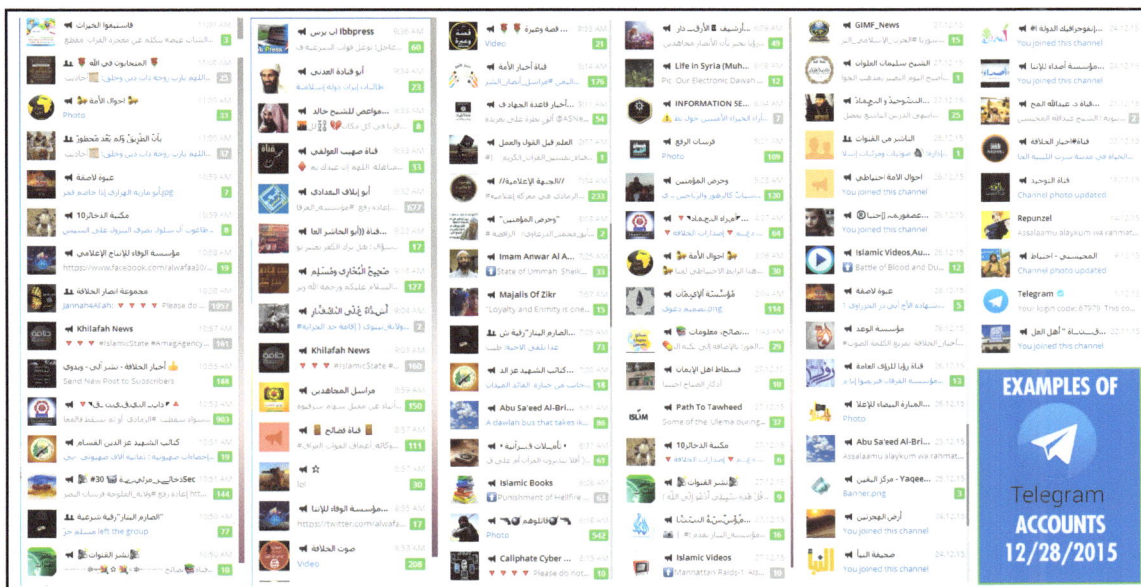

EXAMPLES OF Telegram ACCOUNTS 12/28/2015

Since jihadis began using it last fall, Telegram gained in popularity with its November 2015 release of a desktop version, which made it more like Twitter. Also contributing to the surge in Telegram's use by jihadis is the fact that they have become more adept at it, utilizing its encryption and privacy options once they are established on the platform. The ease of using Telegram was underlined in an article recently posted on a jihadi Telegram channel, which pointed out that on Telegram, in contrast to Twitter, "[there are] fewer suspensions, [material is] easier to publish, and reporting [of channels] is limited."[6]

One of Telegram's features, the secret chat, was used recently by ISIS supporters to draw up plans to target U.S. and British nationals, as well as actual collaboration in a plan to target the "U.S. embassy in the U.K."[7]

MEMRI has repeatedly noted that in order to solve the problem of jihadi and terrorist content online, including on social media platforms such as Telegram, there is a need for an industry standard set by all social media companies coming together and agreeing to tackle it. Until this happens, jihadi groups and individuals will continue to migrate from platform to platform as measures are taken against them. This includes new platforms, which, when created and when jihadis discover them and begin using them, do not know yet how to deal with this phenomenon. This is what happened, for example, with Twitter – when jihadis first began using it, no action was taken, and it soon developed into a major tool for terrorists, both individuals and groups. The same is happening with Telegram today – and since nothing has been done during this entire time, it is growing into a major challenge for counterterrorism officials in the West.

Telegram And Its Refusal To Take Action Against ISIS And Other Jihadi Accounts

A January 8, 2016 MEMRI report, 'Supporters Of The Islamic State' – Anatomy Of A Private Jihadi Group On The Encrypted 'Telegram' App, Offering Secret Chats And Private Encryption Keys, focused on Telegram and its refusal to take action against ISIS and other jihadi accounts. The November 13, 2015 Paris attacks relaunched the debate about Islamic State (ISIS) and other jihadi use of encryption technology and apps, with particular attention, and unprecedentedly negative media coverage, directed at Telegram, which these groups and individuals now heavily favor. Much of this negative reporting about Telegram came as a result of a previous MEMRI report, Jihadis Shift To Using Secure Communication App Telegram's Channels Service, released two weeks earlier and heavily cited in the media.

Telegram developer Pavel Durov, who previously had consistently refused to remove ISIS and other jihadi groups and channels from the platform, grudgingly posted on November 18, "This week we blocked 78 ISIS-related channels across 12 languages."

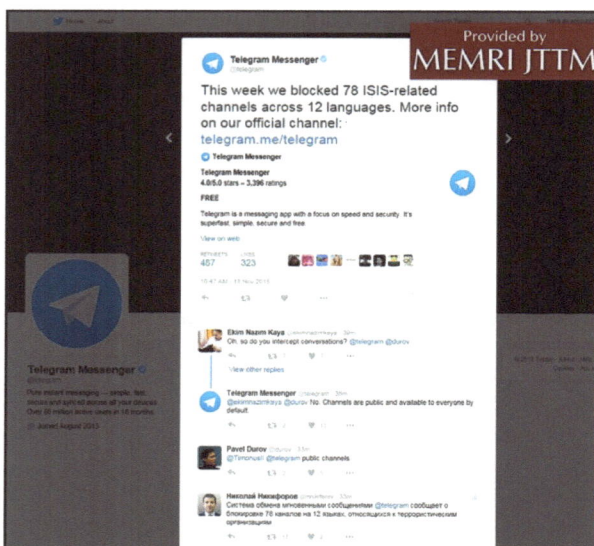

In another tweet, Telegram stated: "We could identify and block these public channels thanks to reports you sent to abuse @ telegram.org. Thank you!" and linked to its FAQ page.

Jihadi accounts shared news of the blocks (see also MEMRI report):

On November 19, 2015, Durov tweeted, including to the MEMRI Twitter account, that "groups are not channels. And we've been against ISIS public content since forever (see our FAQ)."

The same day, MEMRI refuted Durov's statements with a series of tweets to him highlighting how ISIS and Al-Qaeda have embraced Telegram:

MEMRI's tweets to Durov showing ISIS and Al-Qaeda content on Telegram

In an interview following Durov's announcement that Telegram was shutting down groups, MEMRI Executive Director Steven Stalinsky explained to *The Washington Times* how ISIS was using Telegram and predicted that even if ISIS-connected groups were removed, they would not be gone for long and would very quickly be back. This turned out to be exactly what happened – ISIS and Al-Qaeda groups and accounts, notably Nashir, Fursan Al-Raf, CyberCaliphate, Al-Battar, and Global Islamic Media Front (GIMF) – announced shortly thereafter that they had returned to Telegram. Stalinsky said: "We are sure there are lots of private encrypted discussions happening that are just not public but continue... It's positive he [Durov] removed the accounts, but he did it as a temporary thing so as not to get bad press and pressure."

Durov's claim that Telegram "has been against ISIS public content since forever" does not reflect the reality of how the group continues to use the service freely.[8]

The examples of jihadis' use of Telegram in this report constitute only a small part of the research being carried out by the MEMRI Jihad and Terrorism Threat Monitor (JTTM) and Cyber Jihad Lab (CJL) research teams. These teams are following these and other, similar accounts 24/7, due to the sheer number of jihadi Telegram accounts and the volume of content that they are disseminating.

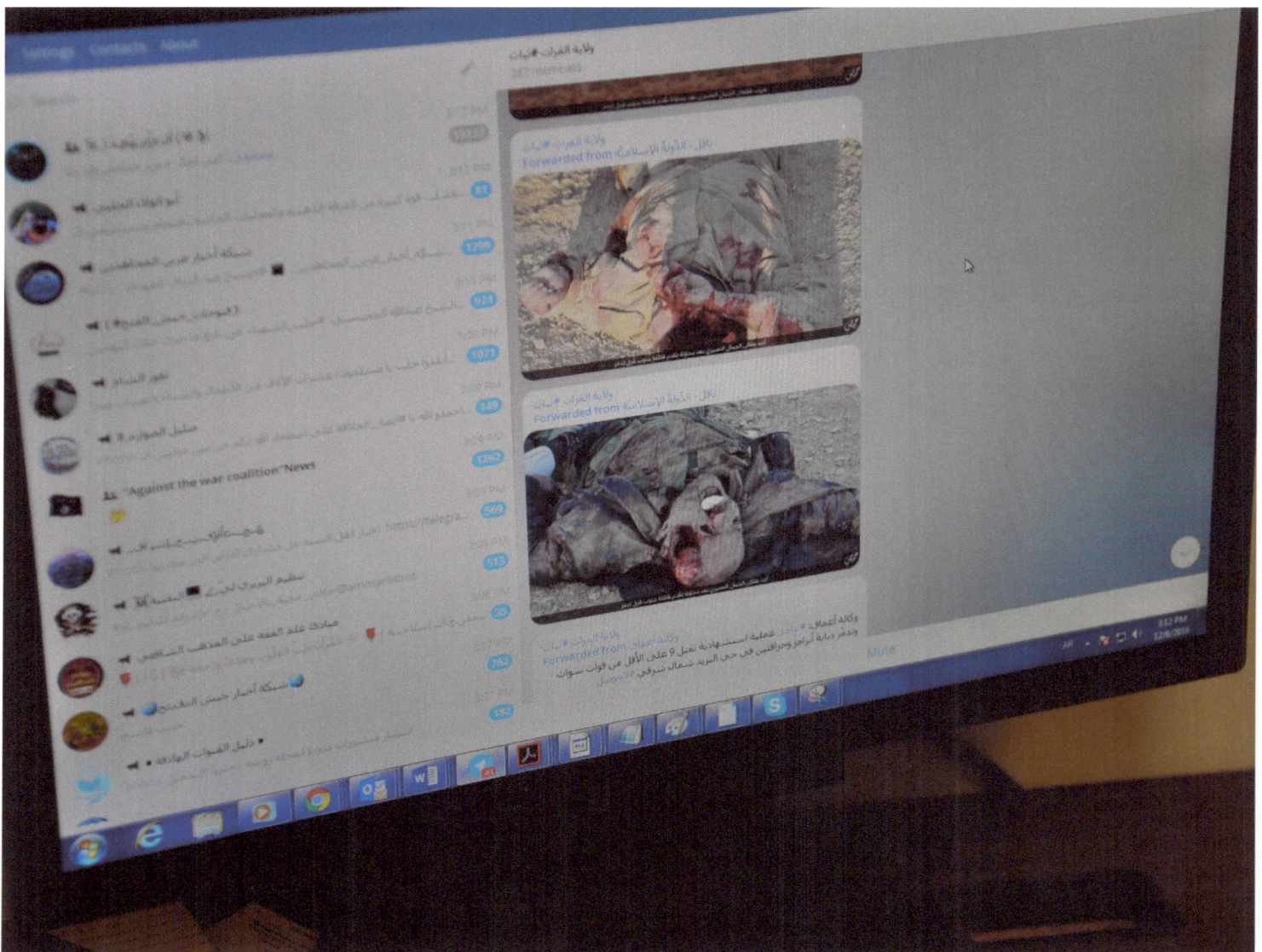

How Jihadis Are Using Telegram: Recruiting, Encouraging Lone Wolf Attacks, Releasing Official Content

Highlighting the popularity of Telegram among jihadis, ISIS released the 30-minute audio recording of the most recent speech by leader Abu Bakr Al-Baghdadi via the platform (see MEMRI JTTM report ISIS Leader Al-Baghdadi Responds To Military Campaign To Retake Mosul, Urges ISIS Soldiers To Remain Strong, Calls For Attacks On ISIS's Enemies, November 2, 2016). Jihadis today are using Telegram for recruitment, outreach, posting announcements, distributing content including guides and advice for lone-wolf attacks and manuals for bomb-making and other terror activity, disseminating the findings of hacking attacks including "kill lists," planning attacks and identifying possible targets, issuing threats, claiming responsibility for attacks, posting pledges of allegiance to jihadi groups, fundraising, and more – including posting videos and photos of executions, crucifixions, and beheadings.

Some examples of jihadi recruitment on Telegram include outreach to engineers and scientists to collaborate, also via the platform, on military projects for ISIS, and recruitment of women. Al-Qaeda's main media wing Al-Sahab recently joined many other jihadi groups on Telegram by launching an official channel, and the content posted on it has included videos of Al-Qaeda leader Ayman Al-Zawahiri, such as his message marking the 15th anniversary of the 9/11 attacks. It has even been used by groups, such as Muslims Safety Tips, to warn users not to trust other apps – for example, in April 2016 jihadis circulated warnings on Telegram about WhatsApp's newly launched encryption, claiming that it was easily hacked and also was untrustworthy.[9] In late August, ISIS announced on Telegram the death of its spokesman and strategist Abu Muhammad Al-ᴬdnani.

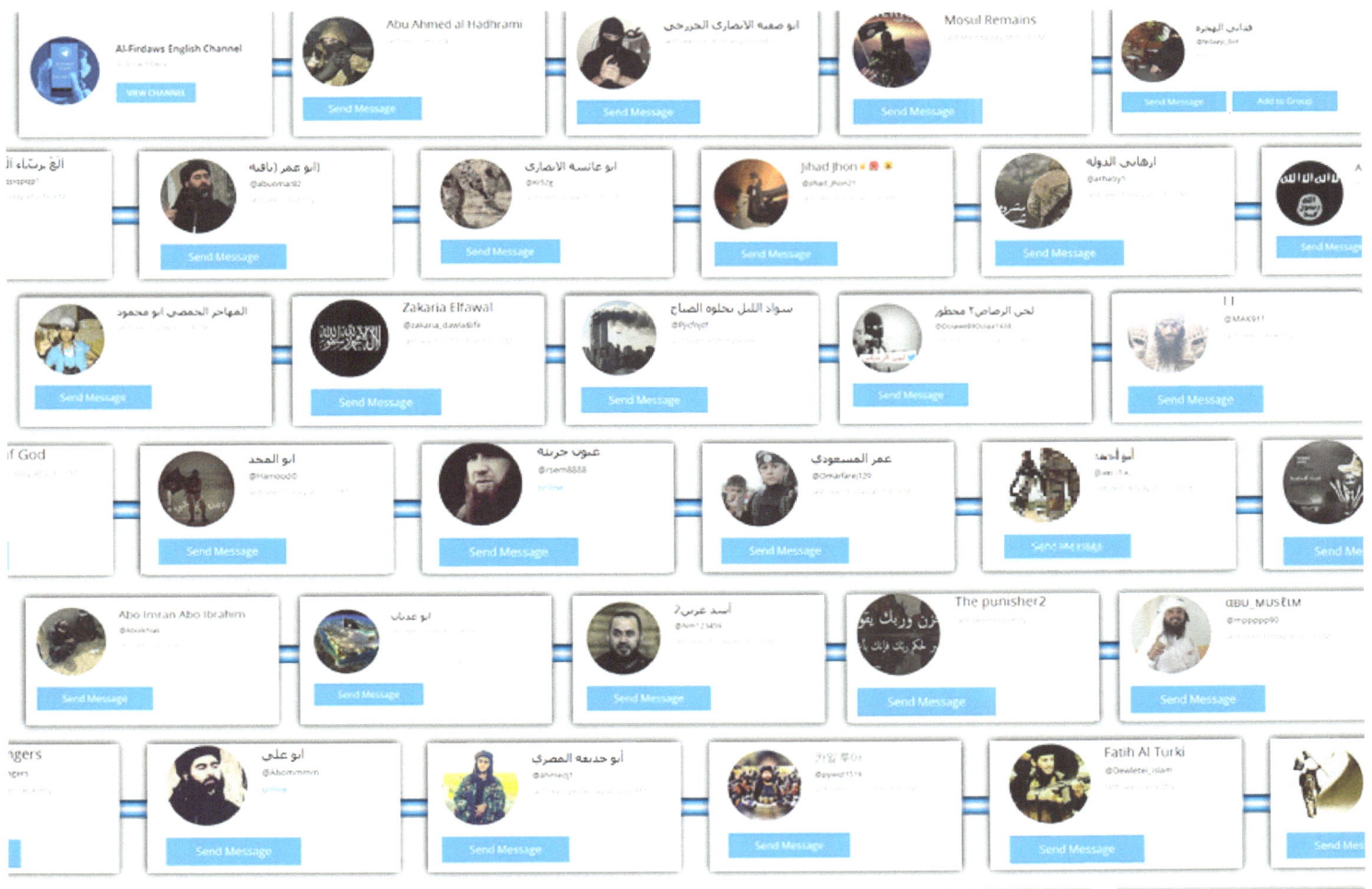

Pro-ISIS groups have also recently shared guides and instructions for lone-wolf attackers; one group distributed a "scorecard" of stabbing, bombing, and vehicular attacks checked off and shooting, poisoning, and beating still unchecked.

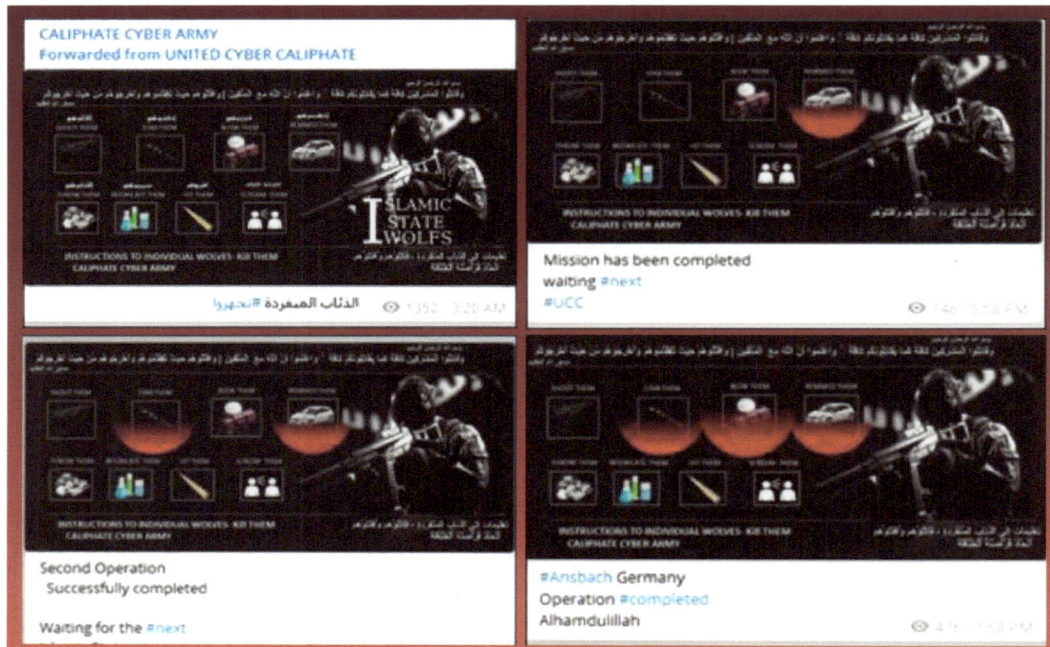

Pro-ISIS Caliphate Cyber Army hacking group updates its "lone wolf scorecard" marking stabbing, bombing, vehicular attacks – with shooting, poisoning, and beatings not yet checked off.

"Kill lists" distributed this past summer by ISIS and pro-ISIS groups on Telegram have included names, addresses, and other personal details, obtained through hacking, of hundreds of: U.S. State Department personnel, U.S. Air Force and Army personnel,[10] police officers in New York and New Jersey, employees of major U.S. companies such as Microsoft, Wal-Mart, and ExxonMobil, Canadian citizens, people labeled "Crusaders and Jews," Americans in Texas and New York, and many more. In addition to "kill lists," other hacking posts on Telegram have included threats of a hack of a SCADA (Supervisory Control and Data Acquisition) system, hinting at London's power grid; an announcement of a hack of 5,000 Twitter accounts in response to Twitter's crackdown on jihadi content on its platform; and boasts and claims of hacks of a Saudi government portal and government database, and of Facebook and Twitter accounts.

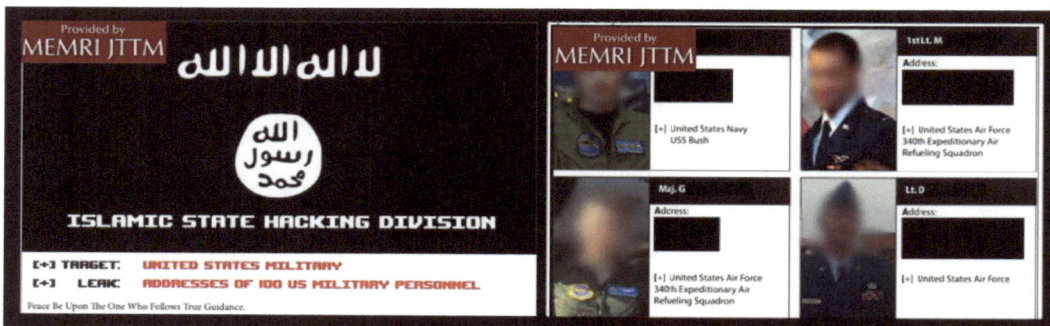

"Kill list" released by a group called Islamic State Hacking Division. See MEMRI JTTM report ISIS Supporters Publish Hit List Of 100 US Military Personnel, With Pictures And Addresses, March 22, 2015.

ISIS operatives in two attacks in France this summer – one against two police officers and the other in a church in Normandy, during which they slit the throat of a Catholic priest – had communicated with each other on Telegram, and Telegram was used in two other thwarted cases in France, one involving a 16-year-old girl.[11] The Normandy attackers also disseminated an oath of loyalty to ISIS via Telegram, as did a Brazilian jihadi group.[12] Threats of attacks circulated this summer on Telegram have included a pro-ISIS media outlet's threats against gay pride parades and the posting of satellite images of U.S. military bases for possible targeting.

Among the claims of responsibility for attacks was ISIS's claim of the July 18 axe attack aboard a train in Germany, and its claim that it downed a Russian helicopter in Syria. In early September, ISIS claimed responsibility for attacks in Copenhagen on Telegram.

There are numerous indications that Telegram is continuing to surge in popularity among jihadis. It hosts jihadi accounts in Arabic, English, Spanish, Portuguese, Italian, French, Russian, Farsi, and Asian languages, and new languages are being added all the time. Prominent jihadis and leading sheikhs once on Twitter are now on Telegram, either in addition to or instead of Twitter. For example, in February 2016, top ISIS propagandist "Turjiman Al-Asawirti," who operates his own pro-ISIS "media production company," launched a channel on Telegram.[13]

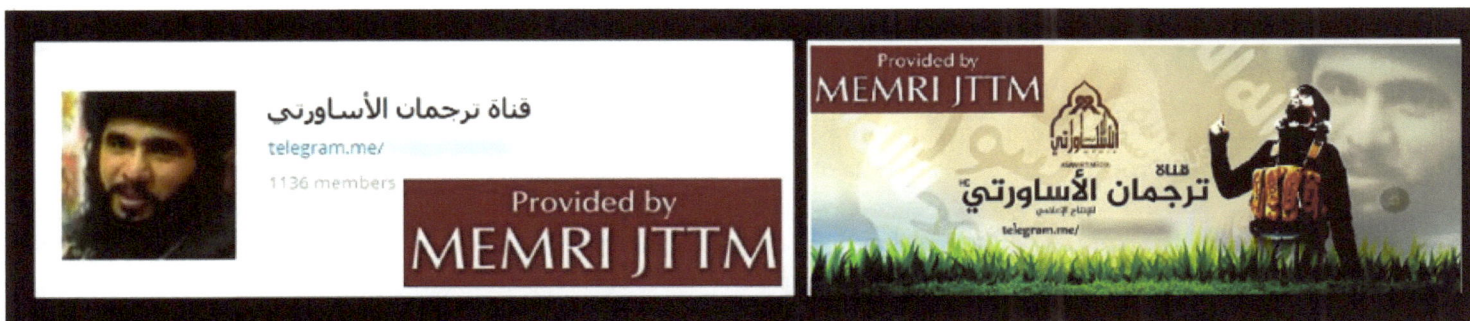

Turjiman Al-Asawirti's Telegram channel

Thousands of new jihadi accounts belonging to groups and individuals are being created weekly, adding to the jihadi groups already operating on the platform – ISIS, Al-Qaeda, Al-Qaeda in the Arabian Peninsula (AQAP), Jabhat Al-Nusra/Jabhat Fath Al-Sham, Taliban, Hamas, Turkestan Islamic Party (TIP), and more.

Some jihadis, such as ISIS's top Southeast Asian propagandist Bahrun Naim, rely on Telegram features such as its private channels and "bot" function, which allows mass dissemination of messages and other types of content. According to counterterrorism experts, Naim is the chief propagandist for Katibah Nusantara – a brigade of Malay and Indonesian fighters in Syria and Iraq. One Telegram bot that features Naim's image on its profile greets new followers with an automated message in Indonesian: "The writer's name is Bahrun Naim but he is commonly known as BN. His full name is Muhammad Bahrunnaim Anggih Tantomo. He was born on September 6, 1983." The bot's automated messages include step-by-step instructions on the production of ammonium nitrate, which is used in explosives, using household materials. Another message includes instructions on producing RDX, the main element in C4 explosives, as well as detonators. The bot also sends out propaganda messages and videos, including of interviews with militants and combat footage.

In November 2015, Reuters contacted a man claiming to be Naim on Telegram, who said that he was "just waiting for the right trigger" to "carry out an action" in Indonesia. Two months later, in January 2016, Jakarta was hit by a gun and bomb attack that authorities say was remotely orchestrated by Naim. The attack killed eight, including four of the perpetrators.[14]

About Jihadis' Use Of Telegram And Its Ramifications

About Telegram

Telegram, which describes itself as "a cloud-based mobile and desktop messaging app with a focus on security and speed,"[15] can run on Android, iOS, OS X, Windows, and other operating systems. It provides optional end-to-end encrypted messaging with self-destruct timers, and also allows users to edit messages already sent.[16] A Telegram user must have a phone number, and the number must be verified before the platform can be used; however, experts have noted that it is easier to use a fake telephone number on Telegram than on other messaging platforms such as WhatsApp.[17] It was created by the Russian-born Pavel Durov, known as the "Mark Zuckerberg of Russia,"[18] founder of the popular Facebook-like Russian social network web-site Vkontakte – which has also been used by jihadis;[19] Durov left Russia in 2014.[20] According to its website, the app is avail-able in English, Arabic, Spanish, Italian, German, Korean, and Brazilian Portuguese.

Telegram is so confident that its encryption cannot be breached that has announced contests with large cash prizes for anyone who could crack its encryption; to date, however, no one has managed to claim the prize. The most recent contest offered $300,000, and ended February 4, 2015 with no winner.[21] Experts have stated that it is using "military-grade" encryption.[22]

Telegram states on its website that in order to fulfill its mission "to provide a secure means of communication that work every-where on the planet," and to continue "to do that in the places where it is most needed (and to continue distributing Telegram through the App Store and Google Play)," it must "process legitimate requests to take down illegal public content" but stressed that this "does not apply to local restrictions on freedom of speech." The website continues: "For example, if criticizing the government is illegal in some country, Telegram won't be a part of such politically motivated censorship. This goes against our founders' principles." It also states: "While we do block terrorist (e.g. ISIS-related) bots and channels, we will not block any-body who peacefully expresses alternative opinions."[23] However, it does not state clearly how to do so.

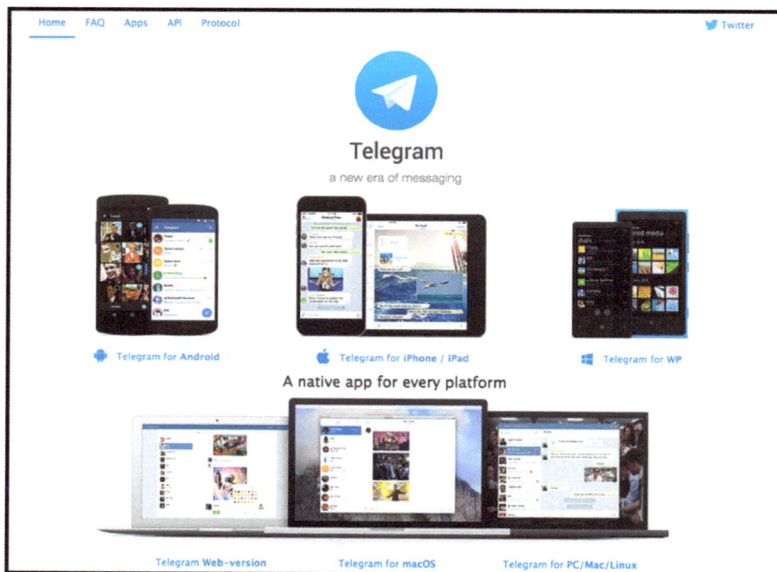

Downloading Telegram on a variety of platforms

Nevertheless, there are many jihadi and terrorist chan-nels on Telegram that have followers or members num-bering in the tens of thousands – from 10,000 to over 80,000.

Some U.S. government officials who are aware of jihadis' use of Telegram, for example, Sen. Angus King (Inde-pendent-ME), say that since the platform is Germany-based, there is little that they can do about it. On Sep-tember 13, 2016, Sen. King noted at a Senate Armed Services Committee hearing on Encryption and Cyber Matters: "[W]e can't stop this. The idea of somehow be-ing able to control encryption is just not realistic... [W]e can deal with Apple or with Microsoft or with Cisco or whoever, but if you've got a cloud-based app that's – the headquarters is in Berlin, and who knows where the data is... there are places our power doesn't reach. We can't regulate something that's over in Berlin or Swaziland."[24]

However, Telegram can be downloaded from the Google Play Store, iTunes App Store, and its website, Telegram.org. It is available for Android, iOS, Windows Phone, Web, MacOS, and PC/Mac/Linux. All of these indirectly share responsibility for the content on the platform.

REASONS JIHADIS ARE SWITCHING TO TELEGRAM

- It offers "secret chats"[25] with end-to-end "military-grade"[26] encryption.

- Secret chats are device specific and are not stored anywhere else – not even Telegram can access them.[27]

- It allows users to "lock" their app with an additional passcode.[28]

- It is user friendly – since ISIS and others began using it about a year ago, Telegram itself has upgraded its platform to be more user friendly, and more like Twitter in many ways, including, in November 2015, introducing a desktop version.

- Users can time their Secret Chat messages to self-destruct after being viewed.

- Anyone can create an account, but it is easier to use a fake telephone number on Telegram than on other messaging platforms.[29]

- Users can send messages, photos, videos and files of any type (doc, zip, mp3, etc)

- Users can create groups for up to 5,000 people, or channels for broadcasting to unlimited audiences, and can make these private, encrypted, and invitation-only. Some of these jihadi groups and channels have tens of thousands of followers/members – up to 82,000.

- Users can reach out to all their phone contacts and also search for other uses by usernames.

- There is less monitoring and suspension of accounts on Telegram than on other social media platforms such as Facebook and Twitter.

U.S., European Officials Sound The Alarm About Telegram

Telegram has been criticized by government officials, both in the U.S. and internationally, for allowing ISIS and other jihadis to operate on it. Following the November 2015 Paris attacks, Rep. Michael McCaul, House Homeland Security Committee Chair, said: "When we saw the encrypted apps on the Paris attackers' iPhone – it was Telegram. When eight attackers and numerous co-conspirators, foreign fighters from Syria, can do something like that and it's completely under the radar screen. We know why it went undetected. It went undetected because they were communicating in the dark space. In a space where we can't shine a light on to see these communications even if we have a court order."[30]

On August 23, 2016, it was reported that France and Germany were pushing for Europe-wide rules requiring the makers of encrypted messaging apps, including Telegram, to help governments monitor communications among suspected extremists. French Interior Minister Bernard Cazeneuve said that French authorities had detained three people with "clear attack plans," but that police needed better tools to monitor encrypted text conversations using powers used to wiretap phones. At a news conference with German Interior Minister Thomas de Maiziere, Cazeneuve said that French investigators, armed with a court order, had been unable to even contact "an interlocutor" at Telegram; Durov wrote, in response to a question submitted via his platform about Cazeneuve's statement, "We haven't received any such request and have no idea what the French officials are after. In any case, Telegram Secret Chats and information on them are not logged on our servers."[31]

Ministère de l'Intérieur 23 août 2016

Ministère de l'Inté 23 août 2016

Also at the news conference, Cazeneuve and de Maiziere insisted that they wanted to work with companies offering encrypted services to ensure that they can't be abused by militants and that they will provide investigators with access to encrypted messages when needed. Cazeneuve noted that Telegram had been used by the killers of the French priest in Normandy in July 2016.[32] Earlier, in March 2016, Paris head prosecutor Francois Molins told 60 Minutes that investigations had "very often" run into a brick wall with Telegram: "We can't penetrate it, we can't get into it."[33] Additionally, Mounir Mahjoubi, president of the National Digital Council, an independent advisory group in France established by former French president Nicolas Sarkozy that focuses on privacy issues, said: "There is an issue with Telegram. They ha[ve] done everything to make it a technological nightmare to find where their server is."[34]

In September 2016, the director of the Dutch General Intelligence and Security Service (AIVD), Rob Bertholee, told the Dutch daily *Volkskrant* that he would like to see limitations on the encryption of apps such as Telegram and WhatsApp, saying that "the threat has not been so great in years." In this, according to the article, the agency is at odds with the Dutch government's position.[35]

TIMELINE OF JIHADIS' USE OF TELEGRAM SEPTEMBER 2015-SEPTEMBER 2016

The following are details of jihadis' use of Telegram and of Durov's statements about it:

- **September 2015:** Telegram launches its channels service

- **September 2015:** Durov acknowledges that he is well aware that ISIS uses Telegram.[36] Asked "Does that concern you?" he answers, "That's a very good question but I think that privacy, ultimately, and our right for privacy is more important than our fear of bad things happening, like terrorism."[37]

- **September 2015:** Durov says Telegram has over 60 million monthly annual users who sent 12 billion messages every day.[38]

- **October 2015:** MEMRI publishes widely cited report on jihadis' shift to Telegram's channel service.[39]

- **October 28, 2015:** The Arab news website Al-Bawaba, reported that 150 terror activists had been arrested by the Egyptian military, after it tracked their coordination using Telegram.[40]

- **November 13-14, 2015:** Just hours after Paris attacks, ISIS uses Telegram to take credit for the attacks.[41]

- **November 18, 2015:** Following ISIS Paris attacks, Telegram announces that it has shut down 78 public channels, in 12 languages, used by ISIS militants or supporters.[42]

- **December 2015:** Officials announce that they believe ISIS Paris attackers had used Telegram to plan, coordinate attacks.[43]

- **November 2015:** Telegram expanded for use on desktops.

- **March 2016:** Paris head prosecutor Francois Molins tells 60 Minutes that investigations had "very often" run into a brick wall with Telegram: "We can't penetrate it, we can't get into it."[44]

- **March 2016:** Durov, in an extensive interview to CBS's 60 Minutes (see below), claims to be "horrified" by terrorist use of his platform but that for him, preserving privacy trumps shutting down terrorism.[45]

- **July 2016:** Al-Qaeda's media wing Al-Sahab begins using Telegram.

- **September 2016:** French Interior Minister Bernard Cazeneuve says, at a news conference with German Interior Minister Thomas de Maiziere about the terrorist threat, that French investigators were not able to contact anyone at Telegram.

2016 Arrests Of Jihadis Worldwide Using Telegram

One of the first arrests linked to jihadi use of Telegram came in December 2014, when alleged ISIS sympathizer Muange Amina Mwaiz, a Kenyan arrested in Hyderabad, India, acknowledged conducting jihadi communications via a Telegram channel.[46] Since then, the number of such arrests has gradually increased. The following are some of the arrests in 2016 of individuals who have used Telegram for jihadi activity:

- January 13, 2016: Australian police raided the home of Sameh Bayda hours after he used an encrypted messaging app, Telegram, on his mobile phone; he was arrested 12 days later after three documents were allegedly found on his phone.[47]

- January 16, 2016: Three Malaysians were arrested by Turkish police after attempting to travel to Syria to join ISIS; according to reports, they had been recruited by Malaysian national Muhammad Wanndy Mohamed Jedi via Facebook and Telegram.[48]

- January 22, 2016: 18 ISIS members were arrested in Bengalu, Hyderabad, India; they were found with explosives and other bomb-making materials. They were found to have formed an organization, Junood-ul-Khilafa – Fil-Hind, aiming to establish a caliphate in India loyal to ISIS, to recruit Muslims to the group and carry out terror attacks in India, at the behest of a Syria-based ISIS operative using social media, including Telegram.[49]

- July 1, 2016: Five suspected ISIS members planning a terror attack in India were arrested; they were in contact via Telegram with the above Syria-based ISIS operative.[50]

- July 23, 2016: 14 people suspected of links to ISIS were arrested in Malaysia; one confessed to sharing information on making IEDs and receiving orders for an attack on Malaysian police via Telegram.[51]

- July 19, 2016: In advance of the Olympic Games in Rio de Janeiro, it was reported that a Brazilian jihadi group, Ansar Al-Khilafah, had used the app to pledge its allegiance to the Islamic State (ISIS). It was also reported that ISIS channels on Telegram were available translated into Spanish and Portuguese.[52]

- August 1, 2016: It was reported that Adel Kermiche, one of the killers of the French priest in the July 26 attack in a Normandy church, had met the other killer, Abdel Malik Petitjean, via Telegram on July 22.[53]

- August 10, 2016: A 16-year-old French girl running a pro-ISIS group on Telegram was arrested and accused of planning a terror attack via the platform.[54]

- August 11, 2016: It was announced that a Filipina had been arrested in Kuwait on suspicion of connection to ISIS; Kuwait authorities said that she had been in touch via Telegram with her Somali husband in Libya, who is alleged to have ISIS connections, and that she had been planning a suicide bombing in Kuwait.[55]

Telegram CEO Durov On Jihadi Use Of His Platform: Denials And Excuses

The following are statements by Telegram CEO Pavel Durov on jihadi use of his platform:

- "I propose banning words. There's evidence [to suggest] that they're being used by terrorists to communicate" – responding to Russian Communication Minister Nikolai Nikiforov's proposal to ban encryption, November 16, 2015.[56]

- "I never mocked the idea of blocking ISIS channels; I mocked the proposal to ban encrypted private messaging due to terrorism" – responding to criticism of Telegram, November 18, 2015.[57]

- "I think the French government is as responsible as ISIS for this [the November Paris attacks], because it is their policies and carelessness which eventually led to the tragedy. They take money away from hardworking people of France with outrageously high taxes and spend them on waging useless wars in the Middle East and on creating parasitic social paradise for North African immigrants." Post on Facebook, November 19, 2015.[58]

- "Since we announced the news, we received dozens of more reports and are blocking confirmed ISIS public channels in real time" – addressing concerns that Telegram is not adequately prepared to address the ISIS presence on it, November 19, 2015.[59]

- "Ultimately... the right for privacy is more important than our fear of bad things happening, like terrorism. If you look at ISIS – yes, there's a war going on in the Middle East. It's a series of tragic events. But ultimately, ISIS will always find a way to communicate within themselves. And if any means of communication turns out to be not secure for them, they'll just switch to another one. So I don't think we are actually taking part in these activities" – in response to a question of whether he "sleeps well at night knowing terrorists use" Telegram, November 20, 2015.[60]

- "Technically it is impossible to deny safe communication only [to] terrorists [and] not jeopardize the personal correspondence of all law-abiding citizens" – explaining Telegram's importance after Apple General Counsel Bruce Sewell called it "malicious and totally unbreakable," March 2, 2016.[61]

- "There's always a risk that your iPhone can be stolen, and the people who stole it can use the data, your private photos, etc, to blackmail you" – explaining why he opposes allowing the FBI to access Apple's database, February 23, 2016.[62]

- "They were probably using other messaging services as well. It's misleading to say that we were responsible – or any tech company is responsible – for that" – explaining why it is not Telegram's fault that it is used by ISIS, February 23, 2016.[63]

- "Never in history have authorities had so much information on their hands as they do today, and they still complain about groups 'going dark'" – dismissing intelligence agencies' concerns about Telegram and other encrypted communications, February 23, 2016.[64]

- "Society in each country has to decide whether they want to make this trade-off, between privacy and what is perceived as high security and lower risk of terrorism" – explaining his support for Apple CEO Tim Cook's decision not to allow the FBI to access the iPhone of San Bernardino attacker, February 23, 2016.[65]

- "Our right for private communication and privacy is more important than the marginal threats that some politicians would like to make us afraid of. If you get rid of emotion for a minute and think about the threat of terrorism statistically, it's not even there. The probability that you will slip on a wet floor in your bathroom and die is a thousand times higher than the probability of you dying as a result of terrorism" – dismissing the threat of terrorism and reiterating his support for private communications, July 4, 2015.[66]

- "You cannot make it safe against criminals and open for governments. It's either secure or not secure. If such a measure [i.e. a backdoor] is implemented, most of our correspondence, our business secrets, our private data would be put at risk. Because if there's a back door, not only a government official could use it, but theoretically criminals including terrorists [could too]"[67] – CNN, February 23, 2016.

- In response to statements by French investigators that they had been unable to contact anyone at Telegram, Durov says "We haven't received any such request and have no idea what the French officials are after." He adds: "In any case, Telegram Secret Chats and information on them are not logged on our servers."[68] – The New York Times, September 5, 2016.

Durov Gives Extensive 60 Minutes *Interview On His Views About ISIS On Telegram*

In an in-depth March 13, 2016 interview with Leslie Stahl on CBS's 60 Minutes, Durov discussed jihadi use of Telegram at length, stating that he is "horrified" by terrorist use of his platform, that if asked by law enforcement he would claim that Telegram cannot unlock any private messages, and that preserving privacy trumps shutting down terrorism. The following are excerpts from the interview:

Stahl: "[Telegram is] also used by terrorists now. Is this a concern for you?"

Durov: "Oh definitely. And in our 100 million users, probably this illegal activity we're discussing are only a fraction of a fraction of a fraction of the potential usage. And still we're trying to, you know, prevent it."

[...]

Stahl: "This was something that you created to allow democracy to flourish, to allow dissidents in Russia and in other countries to communicate with each other. And then all of a sudden you find out that this terrorist group uses your site for completely different reasons."

Durov: "Yeah, we were horrified...

[...]

"But you know there's little you can do because if you allow this tool to be used for good, there will always be some people who would misuse it."

[...]

Stahl: "Is there anything in your mind that says, "Gee, we have to have – to allow law enforcement to get in because what's going on is just unacceptable."

Stahl and Durov, 60 Minutes, March 13, 2016.

Durov: "You know the interesting thing about encryption is that it cannot be secure just for some people.

[...]

"... [T]his is the world of technology and it's impossible to stop them at this point. ISIS could come up with their own messaging solution within a month or so, if they wanted to..."

Narrator: *"Since Paris, Durov has been purging ISIS propaganda from Telegram but says, if asked to unlock any private messages, he would tell the authorities that the encryption code makes it mathematically impossible, using a similar argument as Apple."*

Stahl: "So you're basically saying that even if you wanted to, your hands are tied... You can't do it."

[...]

Durov: "We cannot."

Stahl: "So this is one of the great debates of our time. Which is more important? Is it more important to shut down this kind of terrorism or preserve privacy?"

Durov: "I'm personally for the privacy side. But one thing that should be clear is that you cannot make just one exception for law enforcement without endangering private communications of hundreds of millions of people because encryption is either secure or not." [...]

October 2016: Telegram Claims To Have Department For Processing Complaints About ISIS Usage

In late October 2016, Telegram spokesman Markus Ra said that the company has a dedicated department processing complaints about ISIS usage and that "ISIS channels usually go down within less than 24 hours."[69] However, it is not easy to find out how exactly a user may report such content on the platform. Although Telegram stated in the tweet in which it noted how it had blocked public ISIS channels, "We could identify and block these public channels thanks to reports you sent to abuse @ telegram.org. Thank you!" and linked to a section of its FAQ page,[70] there are no clear instructions at that link, or anywhere else on the page, to do this.[71]

MEMRI RECOMMENDATIONS FOR REMOVING JIHADI CONTENT ON TELEGRAM

Congress: In Washington, at meetings with government officials and on Capitol Hill over the past year, MEMRI has continually reiterated the need for them to call on Germany's Ambassador to the U.S. to testify and answer questions about Telegram. There is also a need to invite founder Pavel Durov to testify as well – and if he refuses, to seek ways to penalize any investments he or his company have in the U.S.

German government: The increase in ISIS threats and attacks should be an impetus for Germany, where Telegram is based, to take action and impose monetary penalties against it and against Durov, and to pressure them to act more responsibly – as U.S. companies, led by Facebook, have recently been doing. MEMRI has also noted that if any jihadi organization was posting kill lists of German government officials, military personnel, and ordinary German citizens, as is happening with Americans, the German government would act immediately.[72]

Google and Apple: Since Google's Play Store and Apple's iTunes App Store are the only outlets for downloading the Telegram app on Android phones and iPhones/iPads, respectively, both Google and Apple are in a unique position to apply leverage to Telegram to remove jihadi content. Furthermore, if safeguards against jihadis using Telegram are not added, then Apple and Android should be asked to remove the app from their stores.

Telegram servers and presence in the U.S.: While not much is known about Telegram's actual presence in the U.S., on October 13, 2016, Telegram tweeted an announcement stating: "Issues in North and Latin America: cooling system died in one data center, massive overheating. Data safe, working with DC staff to fix." Additionally, Telegram states on its page that its servers are "spread worldwide for security and speed." The U.S. government therefore has means at its disposal to penalize it if it does not remove jihadi content, such as fining it or shutting it down here.

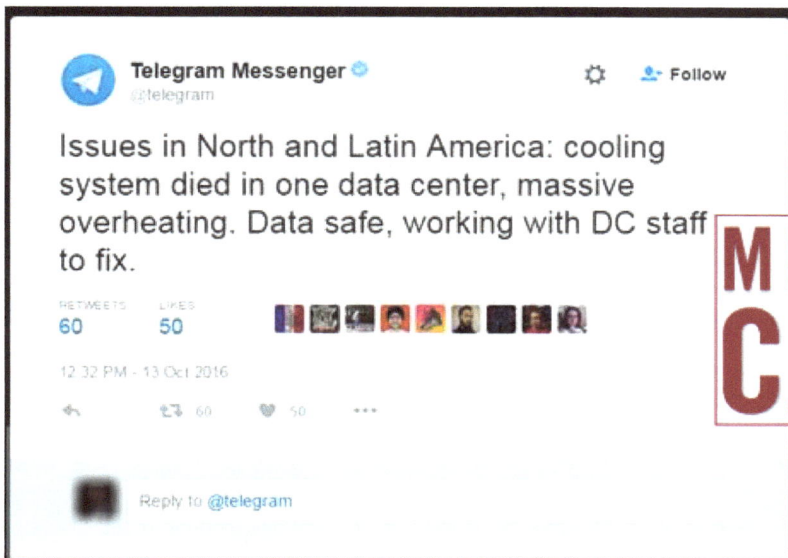

Telegram Messenger
@telegram

Issues in North and Latin America: cooling system died in one data center, massive overheating. Data safe, working with DC staff to fix.

RETWEETS 60 LIKES 50

12:32 PM - 13 Oct 2016

60 50

Reply to @telegram

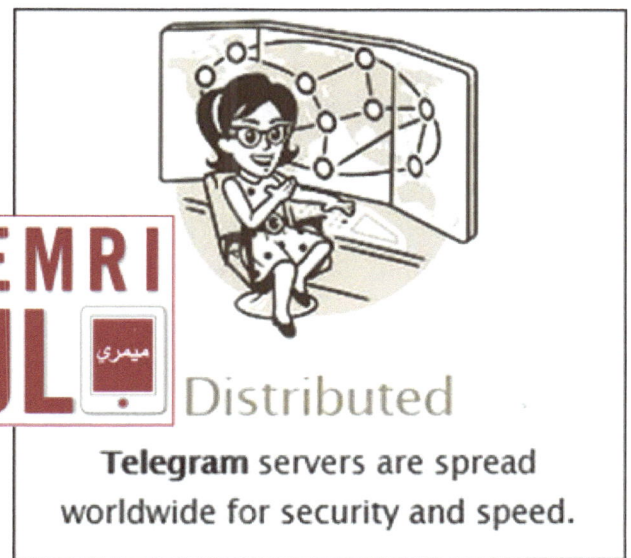

Distributed
Telegram servers are spread worldwide for security and speed.

According to an October 13, 2016 Telegram tweet, the company has "DC staff" and its servers are "spread worldwide."

Steven Stalinsky is Executive Director of MEMRI; R. Sosnow is Head Editor at MEMRI.

GERMANY-BASED ENCRYPTED MESSAGING APP TELEGRAM EMERGES AS JIHADIS' PREFERRED COMMUNICATIONS PLATFORM – PART V OF MEMRI SERIES: ENCRYPTION TECHNOLOGY EMBRACED BY ISIS, AL-QAEDA, OTHER JIHADIS – SEPTEMBER 2015-SEPTEMBER 2016: SECTION 2 – MEMRI RESEARCH DOCUMENTS JIHADI USE OF TELEGRAM

By: Steven Stalinsky, R. Sosnow, M. Al-Hadj, M. Khayat, A. Agron, R. Green, S. Benjamin, and M. Shemesh*

In the past year, the MEMRI Jihad and Terrorism Threat Monitor (JTTM) and MEMRI Cyber Jihad Lab (CJL) have published over 100 reports on the use of Telegram by jihadi organizations, individual members of these organizations, individual supporters, and private channels, in addition to these organizations' media organizations and other news aggregators. As noted, they have used Telegram for recruitment, outreach, posting announcements, distributing content, including guides and advice for lone-wolf attacks, disseminating the findings of hacking attacks including "kill lists," planning attacks and identifying possible targets, issuing threats, claiming responsibility for attacks, posting pledges of allegiance to ISIS, fundraising, and more. The following section focuses on official announcements and releases by various jihadi groups and their media wings, as well as official and unofficial content circulated by these groups and their supporters.

Jihadis Shift To Using Telegram's Private Channels

In the MEMRI JTTM report Jihadis Shift To Using Secure Communication App Telegram's Channels Service, published October 29, 2015, MEMRI Research Fellow M. Khayat noted that since Telegram's announcement of the launch of its channels service, on September 22, 2015, to replace its Broadcast function, numerous jihadis and jihadi organizations had opened their own channels on Telegram. The channels, he wrote, allow individual message content to be transmitted to an unlimited number of subscribed users. At the time of writing, ISIS and Al-Qaeda in the Arabian Peninsula (AQAP) had each created several channels. The new service offered by Telegram constitutes a step up from the standard one-on-one messaging function, and there appears to be no way in to monitor it. Mr. Khayat also noted: "Based on the rate at which new jihadi channels are emerging, and on the large number of members they are attracting, these channels can be expected to become a fertile and secure arena for jihad-related activities." This has indeed come to pass.

As the report noted, "content shared on Telegram channels goes beyond the mere reposting of jihadi groups' propaganda, and includes tutorials on manufacturing weapons and launching cyberattacks, calls for targeted killing and lone-wolf attacks, and more. Some channels, such as those belonging to ISIS, show various levels of coordination among them, even using bots to aid their efforts."

The report stated: "Telegram accounts are tied to the telephone number of the user. A user must be in possession of the telephone with that number that is verified with a code sent to it by SMS or phone call. Telegram also offers a web service, telegram.me, which allows users to open a chat directly with another user by following a URL such as telegram.me/username. Telegram channels use similar URLs, which, in the case of jihadis, are frequently shared and promoted on Twitter and elsewhere. The channels also offer their subscribers a notification function for whenever new content is posted on the page. Since it is a cloud-based app, content is synchronized across a user's devices.

"Telegram channels have some features and functions that are advantageous to the typical jihadi user. First, they provide relative anonymity. A channel displays only the total number of its subscribers to other users without disclosing their names. However, a channel administrator can see the names of members. The relative anonymity makes it harder to identity and track followers of a certain channel for a number of reasons: First, on a social network such as Twitter, following and follower lists are public, and therefore pro-ISIS accounts can be cross-referenced by checking the accounts that they follow and those that follow them. Second, Telegram users can forward content they find on the channel to other Telegram users, thus heightening the sharing and dissemination of jihadi content. Third, messages on the channels are transmitted in a single direction, and no reverse interaction from channel subscribers to the broadcaster is possible. This eliminates the possibility of counter-messaging and the disruption of a content's feed, both of which are used on Twitter as a strategy to counter extremist propaganda. Finally, Telegram provides client-server/server-client encryption as a default option, which, in theory, adds security to the entire interaction." He also notes that Telegram offers a more secure chat option called Secret Chat, which uses end-to-end encryption, and includes features like message self-destruct (see below).

Explaining that Telegram has an application program interface (API) for bots, which are special accounts designed to automatically handle messages, he writes that Telegram bots do not require an additional phone number to set up, and that bot usernames always end in "bot." The Islamic State has a handful of Telegram bots (see "ISIS Bots" below) which aid it in its propaganda efforts. The following are jihadi channels on Telegram included in the report.

ISIS-Related Channels

Nasher Channels
Nasher is currently the flagship of ISIS-related news on Telegram. Nasher was already associated with ISIS when it operated as a webpage (Nasher.me) and as an app as well.[73]

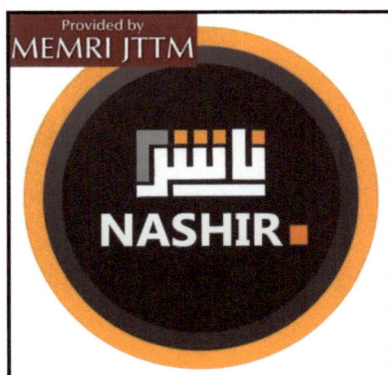

Nasher logo

Nasher operates a number of channels on Telegram that provide ISIS-related news in a number of languages. The main hub of those different channels appears to be the Arabic-language Nasher channel, whereas the remaining channels offer various translations and content summaries. Also, content posted on the Nasher channels mirrors ISIS releases that appear on Twitter and on the Shumoukh Al-Islam forum.

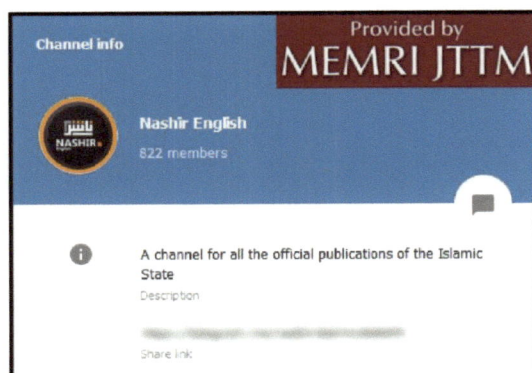

Nasher English channel on Telegram

Nasher channels are offered in the following languages: Arabic (over 10,000 members); English (998 members); French (348 members); Indonesian (1,076 members); German (340 members); Bosnian (201 members); Turkish (181 members), and Bengali (159 members).

Instructions from one of the Nasher channels on how to forward content to other users

a3maqagency

The channel of the ISIS news agency A'maq provides around the clock updates on the Islamic State, focusing primarily on ISIS's military operations. Amaq previously operated on Facebook, but its pages there were constantly being shut down by administrators.[74] Amaq's channel was created on October 1, 2015, and a month later it had over 9,000 members.

ISIS News

A generic channel that publishes ISIS-related news, mostly reposting of various official ISIS releases. At the time of the report's publication in October 2015, the channel had 300 members.

Russian ISIS News

Launched on October 8, 2015, this channel delivers ISIS-related news in Russian. As of October 31, 2015, it had 1,919 members.

KhilafahNews

This ISIS news channel in English had, as of the report's publication in October 2015, 1,711 members.

Promotion of the KhilafaNews channel on pro-ISIS Twitter account (Source: @mhistory087, October 12, 2015)

News3Libya

News for Libya is an unofficial channel reporting about ISIS operations in Libya. The channel had over 500 members in late October 2015.

"Libya and The Honor of the Caliphate"
Also an unofficial channel, reporting on ISIS activities in Libya. The channel publishes various religious content as well. At the time of the report's publication, it had 1,437 members.

Sinai Province
An unofficial channel publishing ISIS news from Sinai. The channel is named after Abu Suhaib Al-Ansari, a slain commander of ISIS in Sinai. The channel had 437 members as of the report's publication, in October 2015.

Fursan Al-Rafi' ("Knights of Upload")
With over 3,000 members, the *Fursan Al-Rafi'* channel is dedicated to uploading ISIS releases on various hosting websites, including YouTube, the Internet Archive (Archive.org), Sendvid, Google Drive, and others.

Al-Battar Media
This channel belongs to the leading pro-ISIS media foundation Al-Battar. At the time of the report's publication, it had over 4,400 members.

Publication Knights Workshop
A channel that focuses on generating pro-ISIS Tweets, which members can readily copy and paste to their Twitter accounts. The channel is affiliated with Al-Battar Media Foundation, and at the time of writing, in October 2015, had 1,205 members.

ISIS Bot

Telegram has several ISIS channels that are run by bots. These channels, which all have a "bot" suffix, provide various functions. It is worth noting that bot channels do not show the number of channel members.

Videodila Bot
This bot will post ISIS releases from its official and affiliated media companies into a user's thread.

"What can this bot do?" Response in Arabic: "Publish releases from [Islamic State] media companies, and media [companies] of [its] provinces, and [releases] from pro[-ISIS media] companies." (Source: Videodila Bot, October 21, 2015)

Al-Qaeda-Affiliated Channels

Al-Qaeda in the Arabian Peninsula (AQAP TV)
The channel, which at the time the report was published had over 2,700 members, circulates releases and statements from Al-Qaeda's branch in Yemen.

Global Islamic Media Front (GIMF)
Created on October 2, 2015, the channel publishes content from the Al-Qaeda affiliated Global Islamic Media Front (GIMF). It had 1,055 members at the time of reporting.

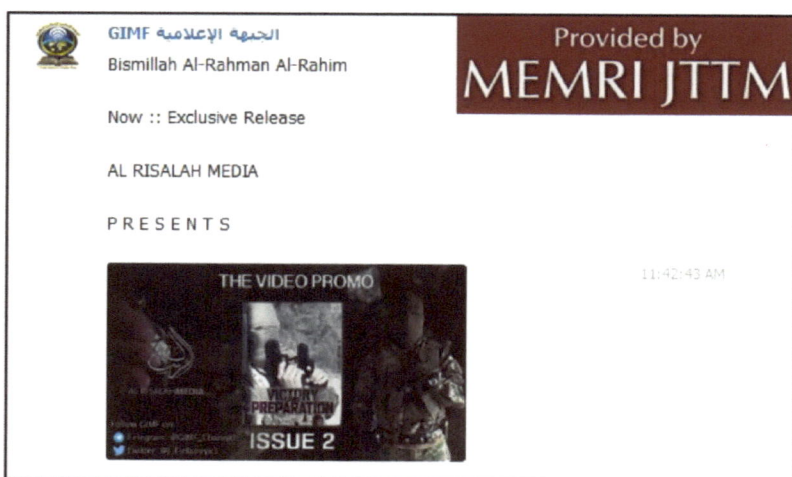

GIMF channel promoting the second issue of the Jabhat Al-Nusra (JN) English-language magazine Al-Risalah

Military Preparation And Incitement

"Be a [Lone] Wolf"
A channel dedicated to sharing military training and weapons manufacturing for the purpose of carrying out attacks. The channel focuses on Saudi Arabia and issues repeated calls to carry out lone wolf attacks there. The channel's motto is "Expel the polytheists from the Arabian Peninsula."

The channel was created on October 18, 2015, and at that time had 51 members. Its channel description reads: "Posting messages and military-related information directed at lone wolves, especially those in the place of [Islam's] revelation [i.e. Saudi Arabia]."

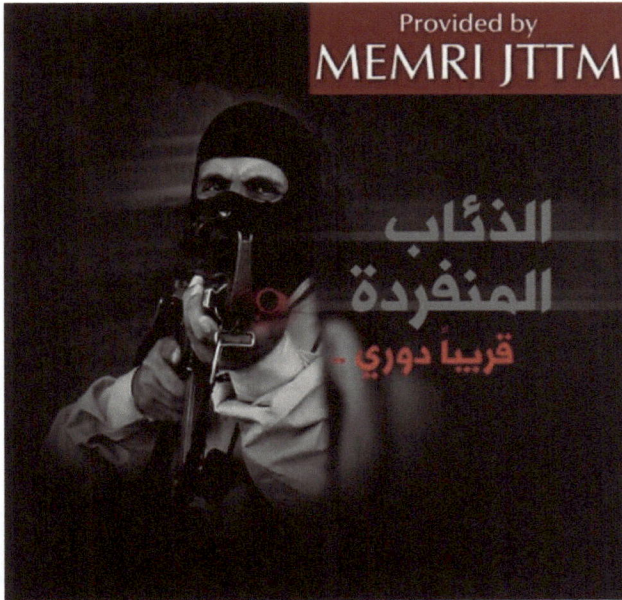

""The channel's profile pic with the caption: "Lone Wolves: Soon [Will Be] My Turn"

Jihadi Designs
The channel publishes jihadi posters and privately created various graphic design. The channel is always on the lookout for new "talent," and also holds contests. It has over 1,100 members. Poster topics vary but include incitement to violence.

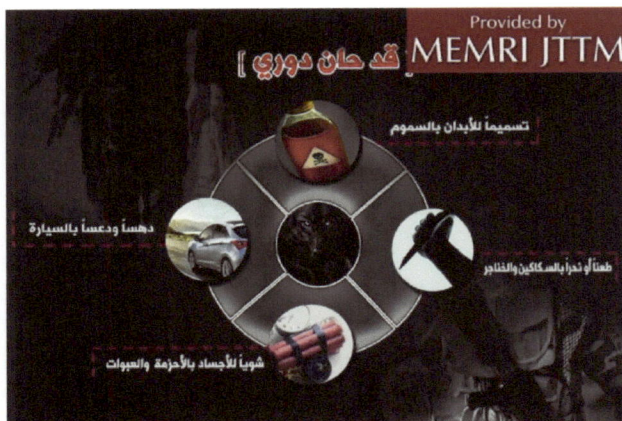

Poster showing various methods of killing people. Top caption in red reads: "It is my turn." Center image is of a wolf, a reference to lone wolf attacks (Source: Jihadi Designs, October 26, 2015)

Jihadis Use Private Groups On Telegram

In the January 8, 2016 MEMRI report "Supporters Of The Islamic State" – Anatomy Of A Private Jihadi Group On Telegram, Offering Secret Chats And Private Encryption Keys, MEMRI Executive Director Steven Stalinsky noted that Telegram offers a number of different options for regular, secret, and encrypted communications, including groups, "supergroups," and channels. Telegram explains these options in its FAQ on its website (telegram.org/faq) as follows: "Telegram groups are ideal for sharing stuff with friends and family or collaboration in small teams, they can have up to 200 members and by default everyone can add new people and edit the name and group photo. If your group grows to a very large community, you can upgrade it into a more centralized supergroup. Supergroups can have up to 1,000 members and have a unified history, deleted messages will disappear for everyone. Channels are a tool for broadcasting public messages to large audiences. In fact, a channel can have an unlimited amount of members. When you post in a channel, the message is signed with the channel's name and photo and not your own."

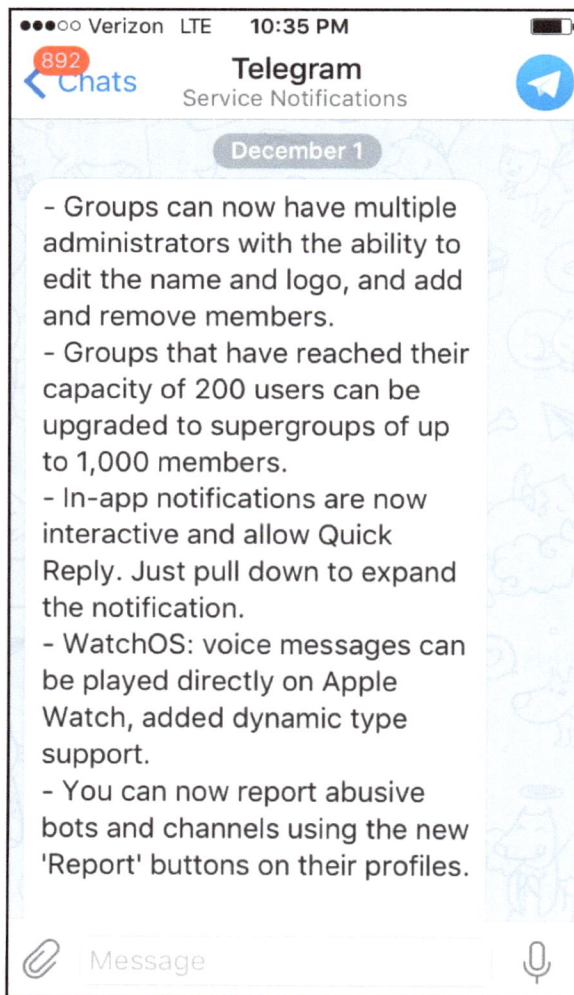

Telegram message to users about groups

"Supporters Of The Islamic State" Group On Telegram

The private ISIS-affiliated group "Supporters of the Islamic State" uses the options available to members: invitations to join the group, private and secret chats, encrypted conversations complete with encryption key, and profile pages of members of the group.

The group, which disseminates ISIS propaganda, was created December 17, 2015 and within two days had nearly 500 members. As part of this group, these members can conduct private conversations with each other or with a small group of other members, and can remain completely invisible to anyone who is not part of the group.

The page for the Supporters of the Islamic State group shows information on its nearly 500 members. The avatars of many of the members include the black ISIS flag and other ISIS symbols (see "Supporters of the Islamic State Group – Member Profiles" section below).

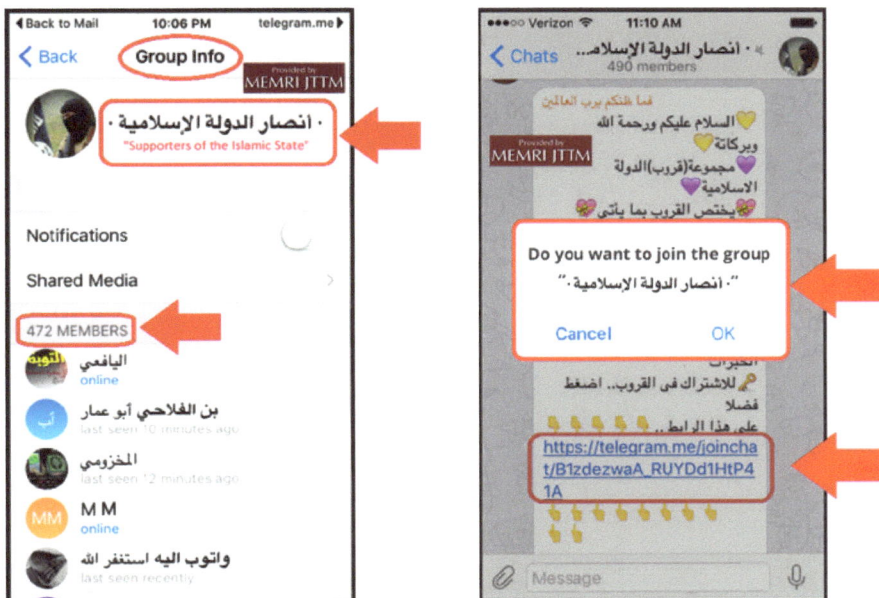

Left: Group page for Supporters of the Islamic State. Right: Invitation to join chat/group

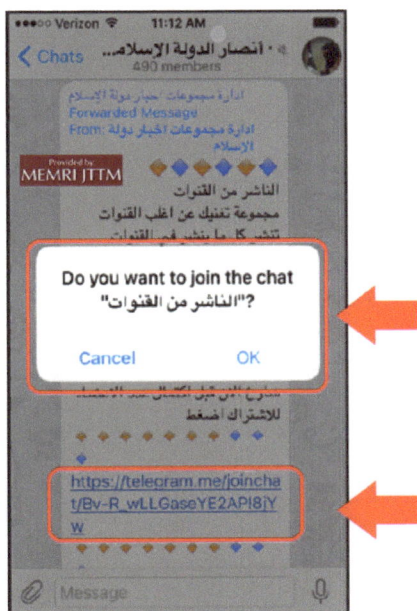

Dialogue box that appears when user clicks on invitation to join chat/group

Secret Chat Feature

Members can invite other members, whether individuals or groups, to participate in a temporary secret chat; these temporary chats are frequently short-lived. Any member of the group can initiate a secret chat with any other member or members by clicking on "Start Secret Chat."

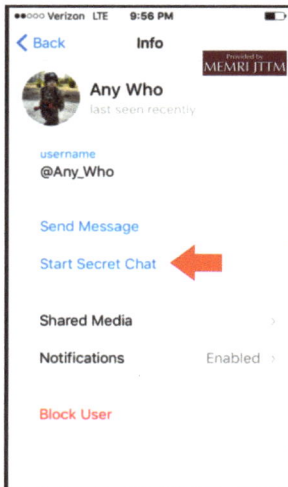

Encrypted profile page features "Start Secret Chat" button

Stand-By Screen For A Secret Chat Window

The dialogue box that appears when a group member invites another to a secret chat explains: "You have invited [other member] to join a secret chat.

"Secret chats:

- · "Use end-to-end encryption
- · "Leave no trace on our servers
- · "Have a self-destruct timer
- · "Do not allow forwarding."

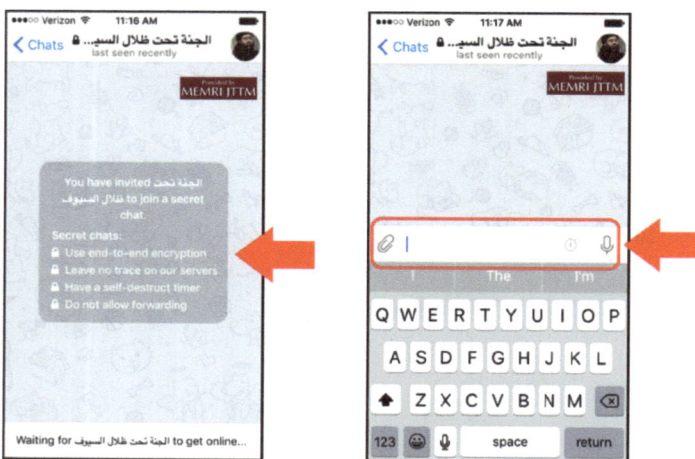

Left: Stand-by screen for a secret chat window. Right: Open secret chat window with blank dialogue box – user can either type or speak.

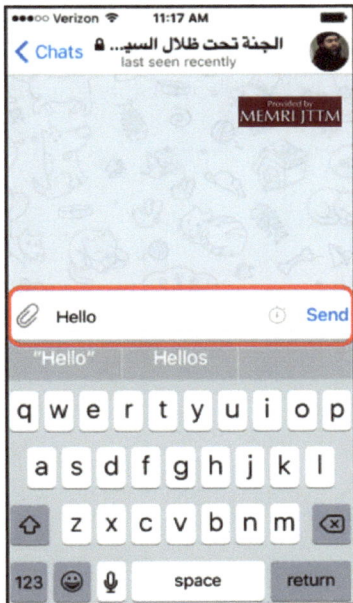

Open secret chat window with message typed into dialogue box; user then hits "send"

Encryption Key For Secret Chats – "200% Secure"

The windows below show a secret chat underway with an "encryption key" (marked in red) that is shared by the chat's participants. The encryption key window states: "This image is a visualization of the encryption key for this secret chat with [interlocutor]. If this image looks the same on [interlocutor's] phone, your chat is 200% secure."

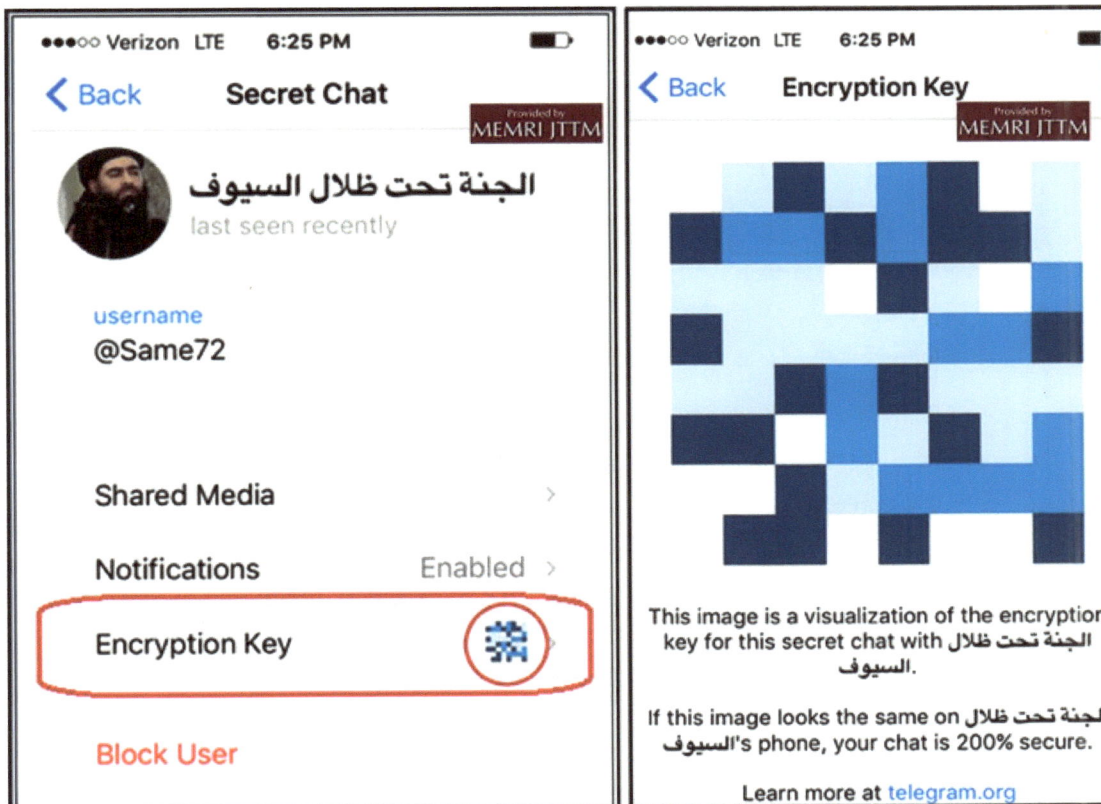

Supporters Of The Islamic State Group – Member Profiles

Following are the profile pages of some of the Telegram users who are members of the Supporters of the Islamic State group:

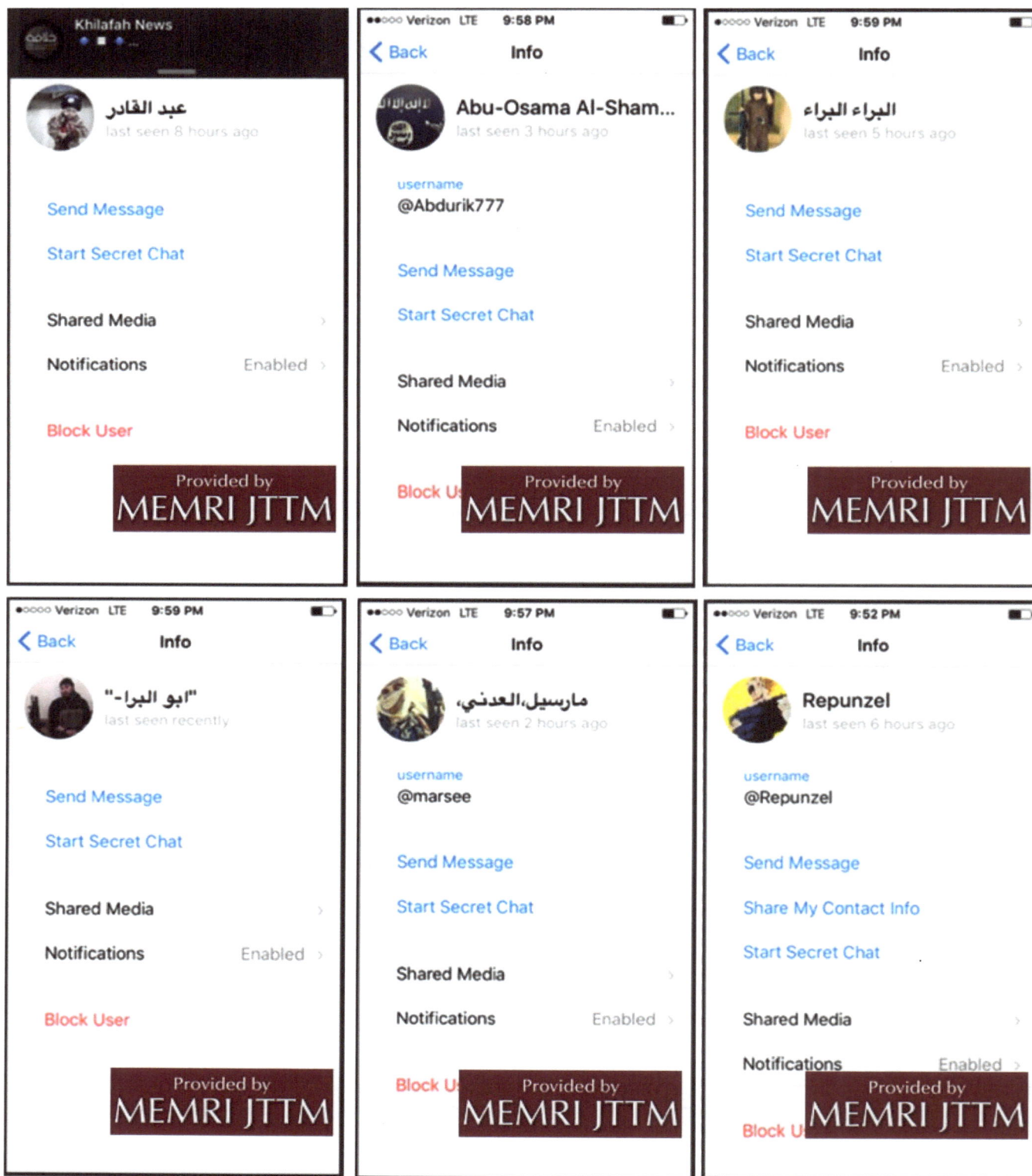

Top French ISIS Recruiter And Executioner Rachid Kassim On Telegram: Promoting, Coordinating, And Encouraging Attacks Via Text, Images, Audio, Video

Rachid Kassim, ISIS fighter, recruiter, media operative, and executioner

In September 2016, it was reported, following the arrest of three teens and a group of young women suspected of plotting jihadist attacks, that French ISIS recruiter and executioner Rachid Kassim, who is currently in Iraq, was believed to be at the center of the plots, coordinating them and recruiting activists primarily over Telegram.[75]

Kassim's online activity has had a direct role in inciting ISIS supporters to carry out attacks in France, such as in the cases of one of the two killers of the French priest in July 2016, Adel Kermiche[76] and killer of a policeman and his partner near Paris in June 2016, Larossi Aballa.[77] A joint investigation by MEMRI French and the MEMRI JTTM was the first to identify Kassim's activity on Telegram.

One of Kassim's channels is "Sabre de Lumière" (Saber of Light), and he also operates another, called "Truth and Proof," which was previously run by Adel Kermiche.[78] After Kermiche's death following the July 26 attack, the channel was inactive for some two weeks. On August 9, it resumed activity with a flurry of messages; by August 18, there were 25 new posts, including text, images, and audio. In one of the posts Kassim wrote: "Thousands of people have access to this list. Continue to disseminate, Inshallah!" During the same period Kassim also posted some 150 new items on his Sabre of Light channel, including 58 audio messages and 48 images. This channel currently has 258 members.

According to police, Aballa was "part of (Kassim's) Telegram group, and Kassim had a real influence in this case," and both French killers, Abdel Malik Petitjean and Adel Kermiche, were influenced and instructed by him, and kept in touch with him via Telegram.[79]

Rachid Kassim's name became known following his appearance in an ISIS video in which he celebrated the July 14, 2016 Nice attack, threatened further attacks in France, and beheaded a prisoner.[80] He has posted several videos in which he called for attacks in France, including a video published prior to the Nice attack in which he suggested using a truck as an operational method (**WARNING – EXTREMELY GRAPHIC IMAGE**).[81]

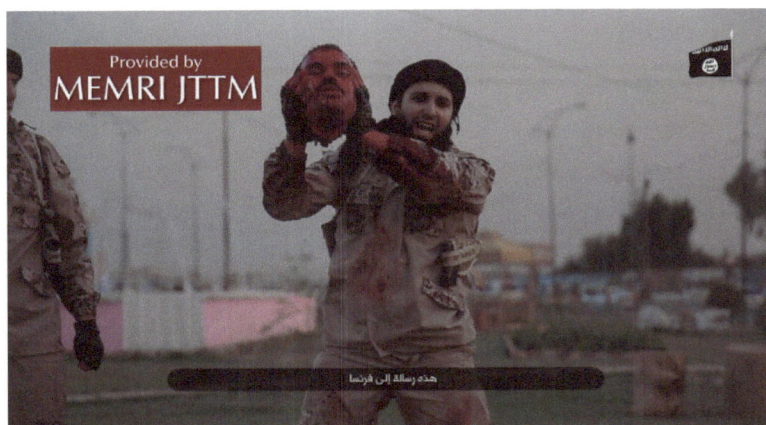

Kassim's posts are mostly religious and doctrinal in character, preaching and promoting the ISIS message and policies, and inciting acts of terror. Evidence shows that he is in contact through Telegram with other ISIS operatives, as well as with ISIS supporters and potential recruits. These communications may indicate involvement in handling ISIS assets in the West, and knowledge of ongoing terror plots.

His main message is a call for "lone wolf" attacks in France and in the West, and he has distributed numerous videos, audios and texts in this vein, including a list of people and places to target.[82] He may or may not be directly involved in the creation of these materials.

Kassim's online activity has increased significantly in the last few months, and he has posted a wide variety of materials including poems by an ISIS member, Q&A with ISIS followers, and reports about martyrdom operations in Iraq and Syria. The theme of sacrifice for the cause is central to his posts, as well as the message that the best option is to be martyred during an attack in the West.

On August 16 he wrote on Truth and Proof: "Give priority to massive attacks... Seven brothers who cannot be detected [by the police] are working on this, so it's only a matter of time". On August 18 he posted on Sabre of Light: "Attack this country of Charlie [reference to *Charlie Hebdo*]. Attack them, do not wait any longer, oh my brothers, in the name of Allah."

In an audio message he posted August 16 on Truth and Proof, he said: "Message to the people of Babylon, the people of Charlie [Hebdo], message to the unjust and criminal people. Soon you will have no more time to go on the Internet, I tell you now. Considering what is going to happen soon, it's really a waste of time to browse Telegram today to see who did what, etc. I advise you to watch your backs because you have been located for some time, but things have been in motion for eight days now, so we wait..." On August 9, on Truth and Proof, he gave a list of potential targets: "A swimming pool, night club, cinema, festivals, theaters... So give priority to massive attacks or targeted attacks on known figures, like journalists or actors... The real heroes of Islam are those who attack in the land of disbelief." The following are examples of Kassim's activity on Telegram, beginning in June 2016:

On June 29, Before Attack In Nice, Kassim Calls For Lone-Wolf Attacks In France: "Go Get A Truck"

In a video released on Telegram, Kassim called for lone-wolf attacks in France, and urged potential attackers to "go get a truck."

He said in a video titled "A Message to France and to My Brothers": "Now, my brother, let us be honest with one another... There aren't many of us here, but there are enough of us for the infidels. Allah be praised, we are facing the beast. We are breaking its teeth, and we hope to chop off its head. But you are in the belly of the beast, my brother. So if you want Islam to be victorious, why would you want to come out of the beast and face its fangs, when you could tear out its heart and its liver? That is what I am asking you. Sheikh Adnani gave you a very clear [message]. He said 'If they close the door on your *hijra*, open the doors of jihad on them, and make them regret it.' Brother, tear up your ticket to Turkey. Allah is right in front of you, brother. In the name of Allah, proceed...

[...]

"Go get a truck... After Sheikh Adnani made his speech, knife attacks were carried out in Palestine. Not many people pointed out this [connection]. A woman took a knife to stab an [Israeli] soldier. In France, you have access to gas tanks, to trucks, and to many products that can cause a disaster, Allah be praised... Just kill five infidels, and end up a martyr. That would be preferable than you coming here and committing martyrdom, killing 100 people.

[...]

"Brother, you are inside the heart of the beast. I ask you: Which victory do you seek? Do you want the victory of Islam, or do you want your own victory?... You know what you have to do, my brother. By Allah, rip out the heart of this beast and tear out its liver..."

Kassim On Telegram: "Attack This Country Of Charlie [Hebdo]," August 22, 2016

On August 18 he posted on Sabre of Light: "Attack this country of Charlie [reference to *Charlie Hebdo*]. Attack them, do not wait any longer, oh my brothers, in the name of Allah." Another post, with photos of the Eiffel Tower and orthodox Jews, stated: "Jihad is not just in Iraq. Jihad is not just in Palestine."[83]

Kassim Calls On Young Algerians To Attack Tourists And Destroy Tourist Attractions, September 22, 2016

On September 21, 2016, Kassim posted a 12-minute audio recording on Telegram in which he called on youths in Algeria, Morocco, Tunisia, and other North African nations to rise up and join the jihad war. He urged young men in these countries to travel to the various battlefronts, such as in Libya, and said that if they cannot do so, then they should wage jihad in their home countries. He began by calling on young Algerians to uproot the idolatry that he said has become common throughout the country, and urged them to destroy statues, icons, and burial sites that attract tourists, and kill those who visit them. He harshly criticized Algerian Muslims who claim to adhere to Islam while living comfortable lives, watching soccer on TV, and stressed that Algeria is *Dar Al-Kuffar* (an abode of unbelief) and that as a result, a jihad war must be waged there against the regime. He also called on them to start small, for example by destroying archaeological sites and statues and killing "sorcerers" and then to attack members of the administration, police, and military, including generals and politicians. He also urged them to film their attacks, and even produce beheading videos, in order to "bring joy to the entire ummah."[84]

Kassim Posts Photos, Names Of Six Belgians, Calling On Muslims To Kill Them, September 26, 2016

On September 26, 2016, Kassim posted on his Telegram account six photos of Belgian nationals, apparently current or former soldiers, along with their names. The photos were taken from their public Facebook profiles. The attached message calls on "Belgian brothers" to kill these individuals. Kassim frequently calls on Muslims in France and Belgium to carry out targeted assassinations.[85] The following are the photos and names:

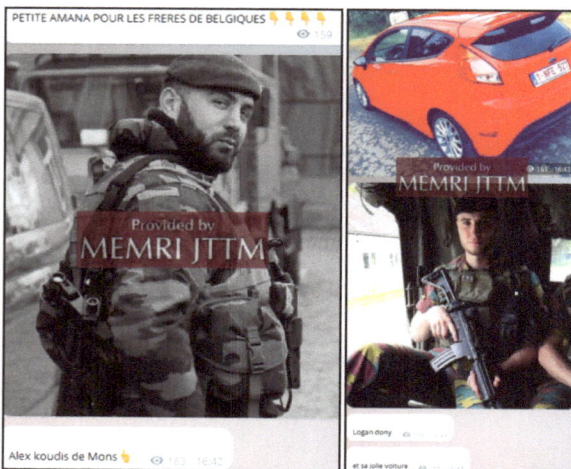

Alex Koudis de Mons; "Logan Dony and his beautiful car"

Kassim Distribute Guides On Using Poison, Making Bombs, September 27, 2016

On September 24, 2016, Kassim[86] distributed a 22-page document in French offering basic instructions on constructing explosive devices, including instructions on processing chemicals to produce several different explosive compounds. It includes a recipe for acetone peroxide or TATP, the explosive most used by ISIS operatives, which was used in both the Paris and Brussels attacks. It also explains how to make detonators, grenades, and mines, as well as a chemical weapon based on chlorine and hydrochloric acid. This kind of bomb-making guide in French is rare, and has not been seen before from any open ISIS sources. Used together with other easily accessible open-source material such as instructional YouTube videos, this material could be used to carry out significant attacks.

According to comments on Kassim's Telegram channel, the booklet was put together by an experienced ISIS operative and is based on translations from the books "Anarchy 'N' Explosives" and "The Terrorist Handbook," as well as on his own personal experiments.

"Making explosives" guide distributed by Kassim on Telegram

List of explosives and where to obtain ingredients to make them

Islamic State (ISIS): Official Releases; Sheikhs, Writers And Operatives; ISIS And Pro-ISIS Channels; Encouraging Lone-Wolf Attacks In The West; Technical Tips For Fighters And Sympathizers; Hacking Groups, Including Release Of "Kill Lists"

Official Content Released By ISIS Provinces, Media Wings, Publications, And More

Chinese And Norwegian ISIS Prisoners For Sale In Issue 11 Of ISIS Magazine *Dabiq*; Available For Purchase Through Telegram, September 10, 2015

Issue 11 of *Dabiq*, ISIS's English-language magazine released on September 9, 2015, contains two pages advertising two prisoners who are for sale. Interested parties are advised to contact a Telegram number that is provided. The first prisoner is a Norwegian named Ole Johan Grimsgaard-Oftadd and the ad for him reads: "To whom it may concern of the crusaders, pagans, and their allies, as well as what are referred to as human 'rights' organizations: This Norwegian prisoner was abandoned by his government, which did not do its utmost to purchase his freedom. Whoever would like to pay for the ransom for his release and transfer can contact the following Telegram number." The ad mentions at the end that this is a limited-time offer. The second page featured displays a Chinese prisoner named Fan Jinghui from Beijing. The same message and personal details are listed for Jinghui.[87]

ISIS Video Sets Out Structure Of Caliphate State, July 6, 2016

On July 6, 2016, ISIS's Al-Furqan media company released a video setting out the structure of the caliphate state. The slickly produced 15-minute video, which was posted on pro-ISIS channels on Telegram, was released in several languages. It details the Islamic State's departments, offices, and committees, but the only official it names is Abu Bakr Al-Baghdadi, ISIS's self-appointed caliph.

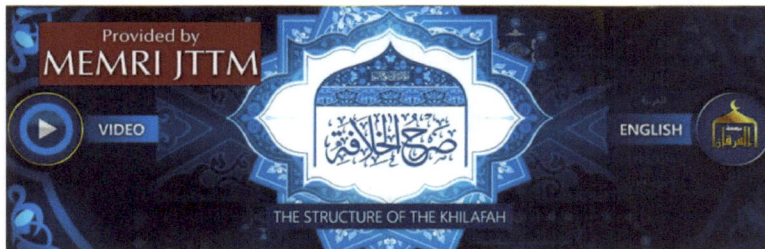

The video recounts the establishment of the caliphate state, the appointment of Abu Bakr Al-Baghdadi as its caliph, and the pledges of allegiance to him that followed; these, it said, came "after years of religious famine and political barrenness." It also says that the caliphate has "outlined the path of salvation and triumph for the Muslim generations." Moving on the to the structure of the caliphate, the video says that it is headed by the caliph, Abu Bakr Al-Baghdadi, who is aided by a Shura Council and a Delegated Committee. Al-Baghdadi is tasked, inter alia, with upholding and spreading Islam, preparing the armies, implementing *hudud* (Koranic punishment), defending the homeland, and fortifying the fronts.[88]

ISIS Announces Death Of Spokesman Abu Muhammad Al-Adnani In Syria, Promises Retaliation, August 30, 2016

On August 30, 2016, A'maq reported that the Islamic State official spokesman, Abu Muhammad Al-ᴬdnani, was killed in Syria. The announcement, that was posted on the agency's website and Telegram channel, cites a "military source," and says: "Military source to A'maq Agency: the martyrdom of Sheikh Abu Muhammad Al-ᴬdnani, spokesman of the Islamic State, during an inspection of his to repel the military campaigns against Aleppo."[89]

In ISIS Video, French Children Threaten: "Today In Syria, Tomorrow In Paris," May 17, 2016

On May 14, 2016, ISIS's Aleppo information bureau released a 14-minute video titled "In the Fathers' Footsteps" (In Arabic), or "In the Footsteps of My Father (in French). The video, distributed via the ISIS channels on Telegram and the Shumoukh Al- Islam forum, focuses on two French boys, Abu Mus'ab and Qaqa, the sons of French ISIS fighter Abu Dujanah Al-Faransi, who remained in Syria after their father died in battle. The video mainly follows the older of the two boys, Abu Mus'ab, who is not yet 12, attending the ISIS school and undergoing military training. Abu Mus'ab is interviewed in the video and talks about the splendid treatment that his family receives from the organization and its fighters. Likewise, throughout the video he threatens vengeance against the West in general and France in particular.

Abu Mus'ab practices his marksmanship by firing at photos of various leaders, including French President Hollande

Abu Mus'ab stands before buildings destroyed by the coalition bombings in the northern Aleppo countryside, and threatens France: "I say to France: Allah willing, we shall kill you, like you killed our Muslim brothers in the land of the Caliphate." In the final part of the video, Abu Mus'ab and his younger brother Qa'qa' execute two condemned men – a Syrian army soldier and a spy. He delivers additional messages to France, noting: "My message to the infidels in general, and to France in particular: You cannot stop us by killing our parents or destroying our homes. On the contrary, this only makes us stronger, and more determined to target you wherever you are." Following the execution, he says: "Today [we kill] in Syria, and tomorrow [we will kill] in Paris, Allah willing."[90]

Abu Mus'ab in class at the ISIS school; Abu Mus'ab and his brother executing Syrian soldiers

ISIS Claims Responsibility For Attack On Train In Germany, July 19, 2016

On July 19, 2016, A'maq published, via its official Telegram channel, a claim of responsibility for that day'sat day's axe attack on a train in Wurzburg, Germany. The claim stated: "A security source [told] A'maq Agency: The perpetrator of the stabbing attack in Germany was one of the Islamic State's fighters. He carried out the attack heeding the calls to target the member countries in the coalition fighting the Islamic State." More than 20 people were wounded in the attack, which was carried out by a 17-year-old Afghan refugee. It should be mentioned that authorities found a hand-made ISIS flag in the teenager's room.[91]

ISIS Claims Responsibility For Suicide Attack In Ansbach, Germany, July 25, 2016

On July 25, 2016, A'maq reported via its Telegram channel that the bomber who blew himself up near a music festival in Ansbach, Germany, wounding at least 12, was "a soldier of the Islamic State." According to the report, the bomber "executed the operation in response to calls to target nations in the coalition fighting the Islamic State."[92]

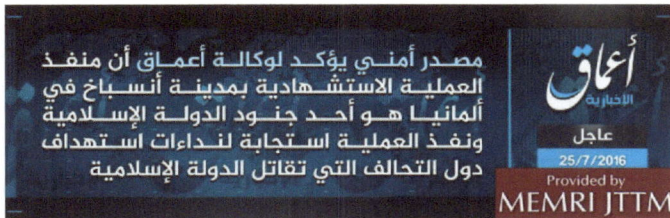

Provided by
MEMRI JTTM

ISIS Releases Video Of Second Normandy Church Attacker, Who Calls On Muslims To Attack France, Coalition Countries, July 28, 2016

On July 28, 2016, A'maq released on its Telegram channel a video message by one of the July 26, 2016 Normandy church attackers, Ibn Omar, whom it identified as Abdel Malik Petitjean. The previous day, the agency had released a video showing the two church attackers, Abu Jaleel Al-Hanafi (Adel Kermiche) and Ibn Omar pledging allegiance to Abu Bakr Al-Baghdadi. Speaking in French, Petitjean addresses Muslims and urges them to attack France: "Strike this country, strike coalition countries, strike them. We strive to support Islam and not ourselves." Repeating his call for more attacks against France, Petitjean says: "I urge all brothers to attack this country. I tell all brothers: Wake up and bring life back to your hearts. Do not listen to the devil who cast doubts, do not be among those who are hesitant, who speak but do not act. Oh brothers, come out, [for] we have everything that we need and we have no excuse. Come out [to attack] with a knife or anything else, and attack them. Kill them en masse, and do not claim that you are ignorant of the truth." The video was subtitled in Arabic.[93]

Provided by
MEMRI JTTM

ISIS Releases Video Of Normandy Church Attackers Pledging Allegiance To ISIS Leader Al-Baghdadi, July 27, 2016

On July 27, 2016, A'maq released on its Telegram channel a one-minute video showing the perpetrators of the previous day's attack on a church in Normandy, France pledging allegiance to ISIS leader Abu Bakr Al-Baghdadi. ISIS claimed responsibility for the attack, in which an elderly priest was murdered. The two men are seen sitting on the stairs inside a house, with one of them holding a printed ISIS banner with "From France" written on it in Arabic. The main speaker says he is Abu Jaleel Al-Hanafi, who is identified in media reports as Adel Kermiche, 19, and names the other man as Ibn Omar. Al-Hanafi then holds Ibn Omar's hand and recites the pledge of allegiance to Al-Baghdadi in broken Arabic. [94]

ISIS Claims Responsibility For Copenhagen Attacks, Says Attacker Responded To Group's Call To Target Coalition Countries, September 2, 2016

Provided by MEMRI JTTM

On September 2, 2016, A'maq claimed responsibility for the shooting attack on two policemen in Copenhagen, Denmark. In a breaking news statement posted on its Telegram channel, A'maq identified the shooter, who was wounded in the attack and subsequently died in hospital, as a "soldier of the Islamic State," stating that he had "carried out the operation in response to the calls to target the coalition countries." Soon after the announcement, ISIS supporters published a photo of the attacker and asked Allah to accept him as a martyr. One post criticized "the crusaders and their agents" for accusing "the soldiers of the caliphate" of being either insane, drug dealers, or mentally ill after each attack saying that they do so in order to save their reputation in their home countries. Promising "the crusaders" more "severe and bitter" attacks in the future, the poste read: "Today is in Copenhagen and we won't tell where it will be tomorrow."[95]

ISIS Claims Responsibility For Attack On Russian Policemen, Releases Video Of Attackers, August 19, 2016

On August 18, 2016, the ISIS news agency A'maq released on its Telegram channel a video of two men claiming that they had carried out the August 17 stabbing attack against two Russian policemen in Balashika, near Moscow, to avenge the Russian bombardments of Muslims in Syria and Iraq. The video identified the two as Uthman Mardalov and Salim Israilov, and described them as "fighters of the Islamic State."[96]

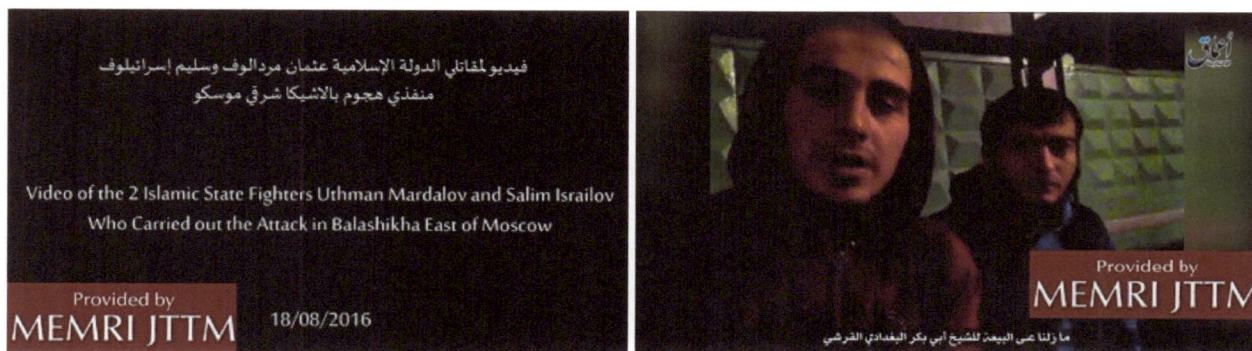

Video of the 2 Islamic State Fighters Uthman Mardalov and Salim Israilov Who Carried out the Attack in Balashikha East of Moscow

Provided by MEMRI JTTM 18/08/2016

Provided by MEMRI JTTM

ISIS Appeals For Support In Philippines, Indonesia, And Malaysia, Calls On Its Supporters There To Attack "Unbelievers" And "Apostates," June 22, 2016

ISIS has appealed for support to Muslims in the Philippines, Indonesia, Malaysia and nearby countries, urging people there to join its ranks and to launch attacks against local "unbelievers" and "apostates." According to a new video by the group that was released on pro-ISIS Telegram channels, ISIS is showing an increase in support for it from various militant factions, namely those operating in the Philippines. The Philippines, it appears, also plays a central role in ISIS's views. The video, for example, names one Abu 'Abdallah Al-Filipini, who was approved by ISIS leader Abu Bakr Al-Baghdadi himself, as the leader of ISIS's cadre of fighters in the country. It also urges ISIS supporters from neighboring countries to make hijra to the Philippines to join the group under the leadership of Al-Filipini, if they cannot go to Syria to join it. The 21-minute video features members from various Filipino militant "brigades," some of which belong to the Abu Sayyaf jihad group, swearing fealty to Al-Baghdadi.

Among the groups named are the Abu Dujana Brigade, the Abu Khubaib Brigade, the Jundullah Brigade, and the Abu Sadr Brigade. Also in the video, ISIS fighters in Syria from Indonesia, Malaysia and the Philippines direct various messages to the group's supporters in their respective countries.[97]

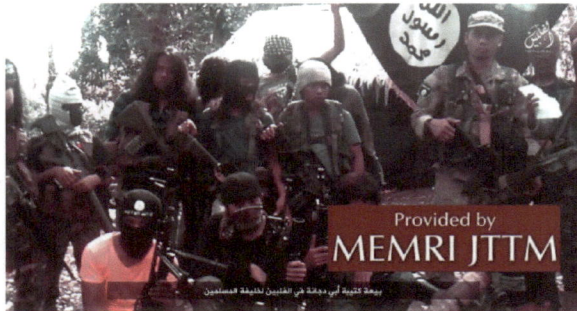

ISIS Infographic Shows Its Areas Of Operation, Boasting Presence In France, Turkey, Many More Countries, June 29, 2016

On June 29, 2016, ISIS released an infographic marking the second anniversary of the declaration of the "Caliphate." The infographic, produced by A'maq and showing the organization's areas of operation, was distributed on pro-ISIS Telegram channels. Three categories of areas of operation are shown. The core areas of Iraq and Syria are presented as areas of "strong control." A second category includes areas where the organization claims to have "medium control": Libya, Nigeria, Egypt, Yemen, Chechnya, Dagestan, Afghanistan, Niger, the Philippines and Somalia. The third category shows areas where ISIS boasts the presence of "security units" (i.e. a network of covert terrorist cells): Algeria, Turkey, Saudi Arabia, Bangladesh, Lebanon, Tunisia and France.[98]

ISIS Video Shows Fighters Deployed In Mosul At Night; Fighter Threatens U.S.: "You Will Be Defeated Again In Iraq And Leave It Humiliated With Your Tail Between Your Legs," October 18, 2016

On October 18, 2016, A'maq published a three-minute video showing ISIS fighters being deployed in Mosul at night. The video, posted on pro-ISIS Telegram channels and elsewhere, was likely released at this time to show that the organization still holds the city despite the campaign to retake it. One of the fighters in the video, who is armed and wears a mask, threatens America and promises that it will once again be defeated and expelled from Iraq humiliated and with its tail between its legs. The video also shows fighters patrolling the streets and securing intersections and streets.

The speaker in the video starts by stating that alongside Allah's promise to grant believers the victory, he also demands that they withstand the test. . He threatens America, stating that it will be defeated, as was foreseen by fallen greats such as Abu Mus'ab Al-Zarqawi, Abu Omar Al-Baghdadi, Abu Hamza Al-Muhajir, Abu Anas Al-Libi, and Abu Muhammad Al-Adnani He adds: "We swear by Allah, swear by Allah, swear by Allah that you, America, will be defeated in Iraq and leave it humiliated with your tail between your legs. Just as our Sheikh Abu Mus'ab Al-Zarqawi vowed and promised when he said you would leave Iraq humiliated."[99]

ISIS And Pro-ISIS Sheikhs, Prominent Writers, Operatives

Top ISIS Propagandist Turjiman Al-Asawirti Launches Telegram Channel, February 10, 2016

On February 10, 2016, top ISIS propagandist "Turjiman Al-Asawirti" launched a channel on Telegram. Al-Asawirti, who operates as his own pro-ISIS "media production company," wrote that his Telegram channel will be used to publish his materials as well as to announce his new accounts on Twitter. Al-Asawirti is among the leading pro-ISIS users on Twitter who have had their accounts repeatedly shut down, yet keep reappearing. Even today, on his Telegram channel, Al-Asawirti announced that he had opened his 419th Twitter account, which was then shut down less than an hour later. He then announced that he had opened his 420th account, which was also shut down shortly thereafter. As of this writing, Al-Asawirti proudly announced his return to Twitter for the 421st time.[100]

Extremist British Preacher Abu Haleema Launches Telegram Channel, Which Is Then Promoted By ISIS Activists, February 29, 2016

On February 28, 2016, the London-based extremist preacher Abu Haleema launched a Telegram channel. Abu Haleema is already very active across various social media platforms, posting excerpts from his videos on Instagram and also sharing, via his official Facebook account, links to his videos on YouTube. Although Abu Haleema himself is not overtly pro-ISIS, his social media followers include ISIS fighters and supporters. He is likely not open about his affiliation due to his May 2015 arrest by Scotland Yard. On his Telegram channel, Abu Haleema stated, "We gonna be exposing these dodgy Taghoot [tyrants] scholars and the kufr [tyrant] and shirk [idolatry] they commit. I will be posting all my vids on here inShaAllah." It should be noted that his Telegram channel has been promoted by other pro-ISIS accounts. As of this date, Haleema's channel had 250 members.[101]

Abu Haleema promoted his Telegram channel on his Facebook account; Caliphate Cyber Army promoted the channel

Top ISIS Writer Asks Twitter, Telegram To Halt Their Suspension Of Jihadi Accounts And Challenges Them, As Well As Anti-ISIS Groups, To Counter-Argue, February 12, 2016

On February 12, 2016, Mu'awiyya Al-Qahtani, aka Ibn Al-Siddiqqah, a prominent writer and ISIS supporter, published an article calling on the founders of Twitter and Telegram to stop suspending pro-ISIS accounts and to instead counter the ISIS message and let people decide for themselves. He invited them to a "challenge and a real confrontation," and called this offer "unbeatable."[102]

Jihadi Writer Warns ISIS Supporters Not To Limit Their Online Activity To Telegram, Urges Them To Use Facebook, Twitter, August 11, 2016

A recent article published by Al-Battar, a media company affiliated with ISIS, warned the group's supporters who are active on social media not to limit their activities to Telegram, and encouraged them to return to Facebook and Twitter. Under the headline "Oh [ISIS] Supporters, Return to the Battlefield," the writer, who identified himself as "Katib," explained that because Telegram does not measure audience impressions, its users cannot gauge how much their messages have spread.[103]

On His Telegram Channel, British ISIS Fighter Appeals For Donations To Keep Fighters Warm In Syria, December 14, 2015

On December 14, 2015, British ISIS fighter Omar Hussain posted a message on his Telegram channel asking for donations to help ISIS militants keep warm in Syria this winter. He wrote: "As we know, winter is soon approaching and the winters here in Sham [Syria] are extremely difficult. We end up having to have 3 blankets wrapped around us at night as some brothers cannot afford heaters. If someone is willing to assist financially in this cause then please message me on my telegram (@Repunzel) and I'll arrange for the money to come inside sham via other secure roots InshaaAllah." This message was viewed by 662 members of Hussain's Telegram channel, which as of this date had a total of 704 members. This is not the first time Hussain has appealed for funds online. On September 24, on his Tumblr account, Hussain explained that militants were in need of financial assistance to purchase military gear, and to fund projects such as building drones.[104]

"Islamic State In Libya" Telegram Channel Spreads "Awareness Of The Expansion" Of The Caliphate In Libya, February 19, 2016

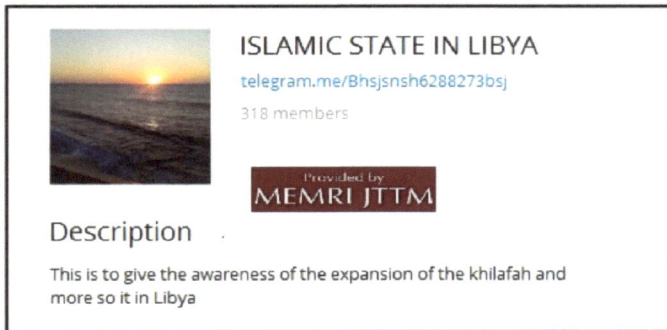

ISLAMIC STATE IN LIBYA
telegram.me/Bhsjsnsh6288273bsj
318 members

Provided by
MEMRI JTTM

Description

This is to give the awareness of the expansion of the khilafah and more so it in Libya

On November 6, 2015, the Telegram channel "Islamic State in Libya" was launched. The channel's description reads: "This is to give the awareness of the expansion of the khilafah and more so it in Libya." The channel, which currently has 318 members, disseminates official photos of ISIS productions filmed in Libya in addition to unofficial personal photos which fighters appear to have taken themselves. The channel also shared the stories of two slain fighters, one from Nepal and another from Nigeria, and a Western fighter calling himself Ali Al Foutawi recounted some of his battle experiences. The photos in the Telegram channel come from official ISIS-sanctioned media productions, and some are also personal snapshots taken by fighters. One photo shows an AK-47 alongside some food items. It appears that the meal was in celebration of an Al-Shabaab leader pledging allegiance to ISIS. The sign in the photo reads: "The Islamic State – We congratulate the pledge of allegiance of our brothers in Somalia – your brothers, Tripoli province."[105]

Pro-ISIS Media Group Releases Video Calling For Attacks In U.S., France, Belgium, Italy, Denmark, Spain, Russia And Iran, August 17, 2016

On August 17, 2016, the pro-ISIS media group Al-Thabat released a video on its Telegram channel urging Muslims to carry out attacks in the West, specifically in the U.S., France, Belgium, Italy, Denmark, and also in Russia and Iran. The video, titled "Come On Rise," begins with a series of slides listing these countries, followed by a narrated message in English, apparently a low-quality translation of a text in Arabic. The message is accompanied by segments from ISIS videos, including many scenes of ISIS beheadings. The video also features excerpts from a recent speech by ISIS spokesman Abu Muhammad Al-Adnani, who is believed to also be the head of the group's external operations, calling for attacks in the West, especially against civilians. The narrator begins by saying that the infidel nations have come together to fight ISIS, and therefore the monotheist Muslims of Europe must rise up and act, for a monotheist on his own is equal to an entire army. He adds: "Raise the banner of jihad in the kuffar's [infidel's] own soil, and fight them the way they fight your brothers. Fight them and make their lives full of terror. Get up and fight them even with a knife... Rise, my unifying [i.e., monotheist] brother, and kill them in markets or in train stations, or in their homes or anywhere... Be on the path of our brother Omar Mateen [perpetrator of the Orlando attack], who shook the throne of the protector of the Cross, America, and also on the path of our brother Larossi Aballa [perpetrator of the attack in Magnanville, France], who arose the *sunnah* of slaughtering in the middle of France..."[106]

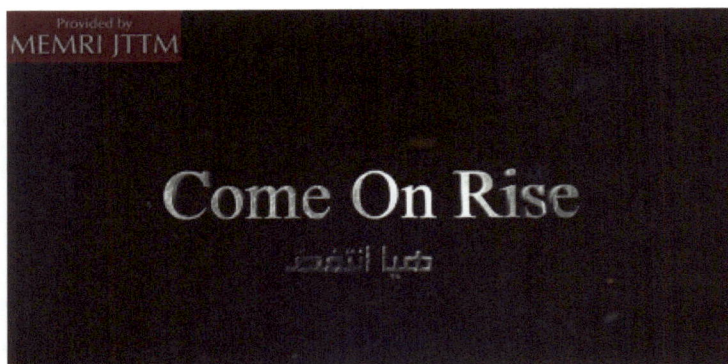

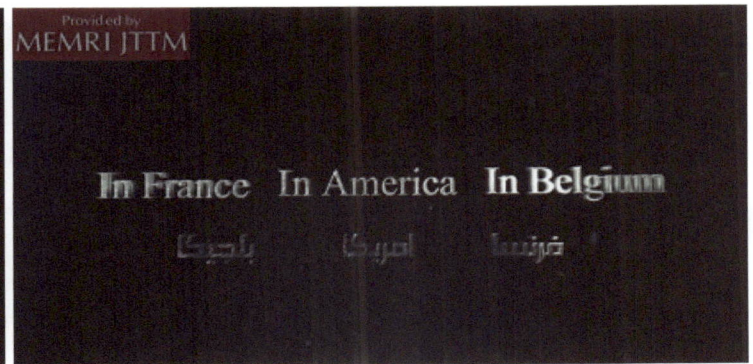

ISIS Activist Promotes His "Online Da'wah Attack Operations" Telegram Channel On Facebook, February 26, 2016

On February 24, 2016, a jihadi Facebook user promoted his new Telegram channel, "Online Da'wah [preaching] Attack Operations," providing a link to it on his Facebook page and explaining its purpose: "Gather amongst us to help in targeting attacks on popular pages/groups with [i.e. by posting on them pro-ISIS comments]." On the channel itself, he posted a link directing readers to a private chat. The Telegram channel states: "Welcome to Online Da'wah Attack Operations Group. We will target groups or pages on social media 2-3 times a week insha'Allah. Please add your brothers and sisters for help in executing these operations with us. Jazakum Allah Khair [may Allah reward you with goodness]."

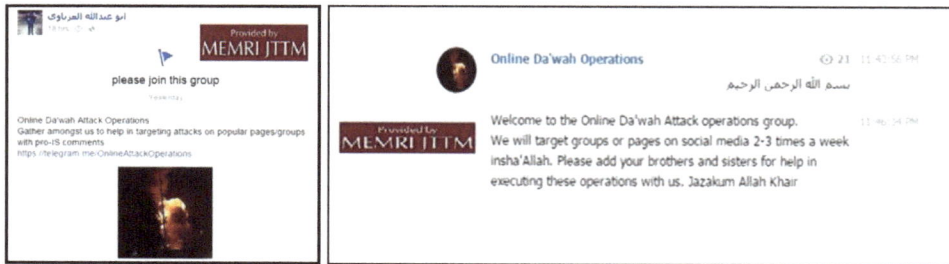

It also says: "Please Ask Friends to Join Group – Gather amongst us to help in executing coordinated attacks on popular pages of kuffar/ Group is 100% pro-ISIS."[107]

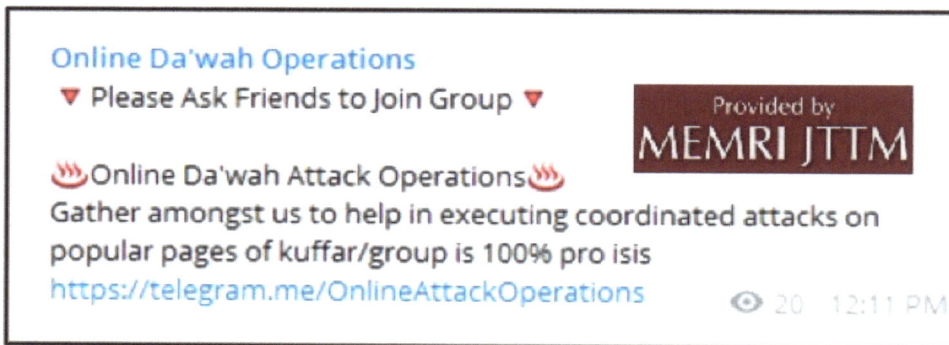

On Telegram, American ISIS Member In Syria Promotes Belgian Recruiter To Her Followers, March 21, 2016

On March 20, 2016, an American woman in Syria who recruits women who might be interested in moving to the Islamic State, has promoted the personal Telegram account of a man who is allegedly a Belgian sheikh who helps interested parties with immigration to the Islamic State and who promotes ISIS ideology. Her husband, also in Syria, is known to help men interested in immigrating to Syria, or in financing projects in the Islamic State; social media accounts frequently refer such interested parties to him. On March 20, 2016, the woman posted: "For Hijrah [immigrating to join ISIS] and dawa'ah [propaganda] advices for Belgium brothers; please contact sheikh Abu Abdulluh Al Beliki."[108]

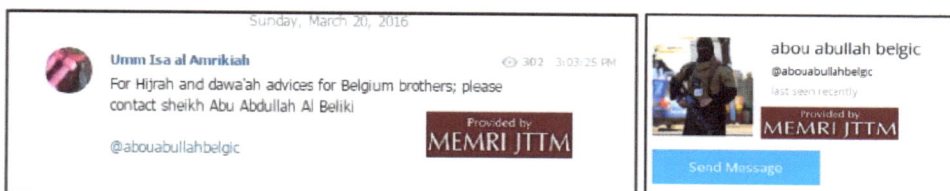

"Shoot Obama" Game Shared In Pro-ISIS Group, April 1, 2016

On March 22, 2016, a user in a private pro-ISIS Telegram chat group shared a link to an Android mobile game called "Shoot Obama." The "Shoot Obama" game files are hosted on Archive.org.[109]

American Who Claims To Be An ISIS Fighter Spreads ISIS Content Via Telegram Group, May 10, 2016

An American claiming to be an ISIS fighter, who calls himself Amriki Muhajer, posts numerous updates about ISIS statements and military efforts, as well as general propaganda.[110]

Female German ISIS Members Recruit On Facebook And Telegram, Share Life Experience Under ISIS Rule, May 24, 2016

A Facebook page and two Telegram channels allegedly operated by female German ISIS members currently residing in Syria and Iraq provide detailed reports about their emigration and life in ISIS held territories. The channels, "The Life of a Muhajira [Female Immigrant]" and "Diary of a Muhajira," regularly distribute ISIS-related media publications and aim to recruit supporters and activists from among German-speaking Muslims. Both channels provide similar content, but appear to be operated by different ISIS media operatives. Both address German-speaking Muslim women living in the West, urging them to join the Islamic State, and offer ISIS supporters interested in immigrating the option to establish contact with them and have their questions answered. By describing the stories of their successful immigration, they aim to serve as examples for others interested in joining ISIS.[111]

Pro-ISIS Telegram Channel Threatens Christians In Tripoli, Lebanon, July 12, 2016

On July 12, 2016, the administrators of a pro-Islamic State Telegram channel called "News of Muslim Lebanon" posted a threat to attack Christians in Lebanon, especially in the northern city of Tripoli. The threat follows a recent series of attacks in the village of Al-Qaa' and the opening of radical Salafi cleric Ahmad Al-Asir's trial.[112]

Pro-ISIS Media Company Celebrates Orlando Attack: "America – The Moment Of Your End Has Come" June 16, 2016

In the days following the June 12, 2016 Orlando nightclub attack, the Al-Wafa' Foundation, a pro-ISIS media company which publishes articles and essays, published several articles celebrating the attack. Touting the fact that it had been carried out during Ramadan, "the month of heroism and victories," the articles expressed the hope that this signaled the beginning of a wave of similar attacks on U.S. soil. They went on to urge "lone wolves" residing in the U.S. to strike in the country, and praised gunman Omar Mateen for answering the call of Allah and the call of ISIS leader Abu Bakr Al-Baghdadi.

Following are excerpts from an article published on Al-Wafa's Telegram channel on June 15, 2016, by one Abu Wahba Al-Gahrib Al-Kinani, titled "To The Taghut Of This Generation, America – The Moment Of Your End Has Come." The article stated that the Orlando attack was revenge for all the invasions carried out by the U.S. in Afghanistan, Iraq, Somalia, and other places.

"This attack by a lone wolf who swore allegiance to the Islamic State was not the beginning, and it will not be the end. And if we examine the yield of this blessed attack against the homosexuals, we will find that during it 50 were killed and over 53 were wounded, Allah and His mercies be praised. All this was carried out first of all, and only, by virtue of Allah the Supreme. All this was carried out by one man – so imagine what would happen if Muslims and monotheists enraged about the desecration of things sacred to Allah united?! By my life, they would shake the foundations. This was also proof of the magnitude of the weakness of the enemies of Allah" which, it added, is expressed also by the U.S.'s fear of launching a land war.

It continued: "This blessed operation came during Ramadan, the month of heroism and victories, and it came in order to differentiate between the miscreants and the good, the hypocrite and the faithful. Whoever now shows solidarity with American and with the homosexuals who were killed never showed solidarity when millions of Muslims were killed in Iraq and are still being killed in Iraq, Syria, and other places."[113]

Swedish ISIS Fighter Launches Pro-ISIS Telegram Channel In Swedish, July 22, 2016

A Swedish ISIS fighter is active on numerous social media platforms. In the past he used Twitter to post messages and photos, but appears to have wearied of having his account repeatedly suspended. He is married to a Malaysian ISIS member. On his English-language Tumblr blog, launched in March 2016, he compares his life today to his previous life in the West, and insists that he is now happier. He also discusses aspects of brotherhood, fallen friends, and the strong bonds of comradeship he has developed with his fellow fighters. In a recent entry, he provided a link to a pro-ISIS Swedish-language Telegram group that he created.[114]

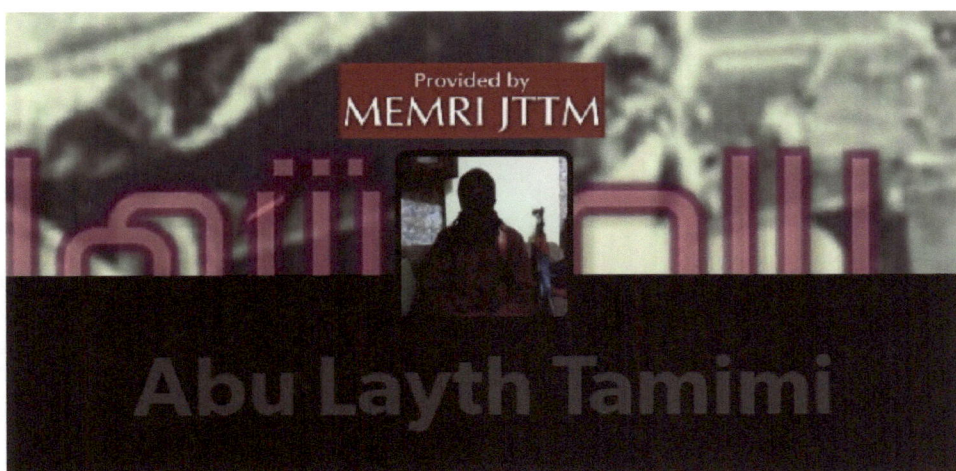

Following Publication Of MEMRI Report About Pro-ISIS French Media Entity, Account Is Shut Down – But Is Relaunched Hours Later, July 21, 2016

On July 15 and July 18, 2016 MEMRI published a two-part report on the French media operation "Ansar At-Tawhid" (AT), active on multiple online platforms including Telegram to spread the ISIS ideology and provide logistical and financial support to ISIS supporters wanting to join the organization or carry out attacks in its name. The reports also pointed to potential links between AT and the French terrorist Larossi Abballa who carried out a double assassination near Paris on June 12, 2016. On the night between June 17 and 18, AT's Telegram account was shut down by the Telegram administration, only to reopen several hours later with the same content and objectives. After the closure of the Telegram account, several other pro-ISIS accounts spread the news and shared the link to the new account:

Below is a screenshot of AT's new Telegram account, with comment by the administrator: "They censor us and we are back 30 seconds later."[115]

Pro-ISIS Telegram Channel Distributes Photos Of U.S. Soldiers At Saudi Base, August 2, 2016

On August 2, 2016, the Arab-language Telegram channel "Exclamation," which posts information about current events, shared photos of what it said are U.S. soldiers training in the Eskan Village base near Riyadh, Saudi Arabia. According to the post on the channel, the photos were taken from social media accounts. Some of the photos appear to be recent, as they show participants wearing shirts from a June 2016 running event in Saudi Arabia. Several photos are from a 2012 running event held at the base. The channel's administrator explained that "Hundreds of American soldiers are present in Eskan Village in the Riyadh airbase, where they operate the Patriot [missile] systems that protect the Saudi skies. They share their daily experiences on social media pages." A couple of the photos were shared by prominent pro-ISIS Telegram channel Dabiq Al-Khilafa.[116]

WARNING – EXTREMELY GRAPHIC: On Telegram, Jihadis Disseminate Death Photos Of Martyrs – Noting Their Beatific Smiles, Scent Of Musk Emanating From Their Bodies, And The Virgins Awaiting Them In Paradise, November 4, 2016

Jihadi groups, including ISIS and Al-Qaeda, have long been sharing photos of dead fighters on social media platforms. Many of these pictured individuals are foreign fighters from countries such as the U.S., U.K., France, Belgium, Germany, Spain, Bosnia, Tunisia, Uzbekistan, and Turkey, who came to Syria, Iraq, and Libya to wage jihad.[117] The following are examples of such posts of martyrs, which can be seen frequently, that have been posted on Telegram in the past few months.[118]

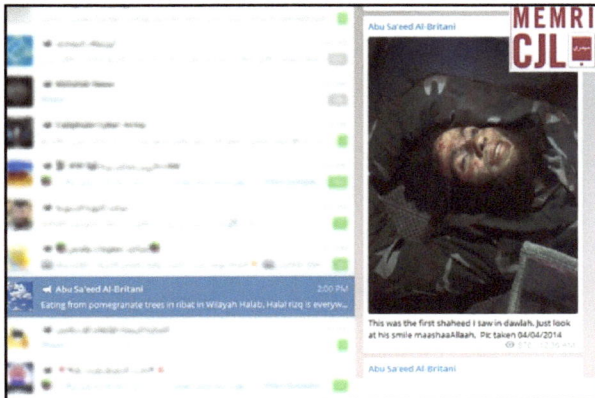

Shared on a jihadi forum on Telegram on December 28, 2015: "Abu Sa'eed Al-Britani: 'This was the first Shaheed I saw in dawlah (the Islamic State). Just look at his smile maashaaAllaah."

Shared on Telegram on August 4, 2016: "Life in Syria" – "A bro reportedly martyred in Haleb [Aleppo]... in sha Allah"

Shared on Telegram on December 23, 2015

Shared on Telegram on January 16, 2016: "He sold [his] life cheap, [but] by Allah... he purchased [Paradise]..."

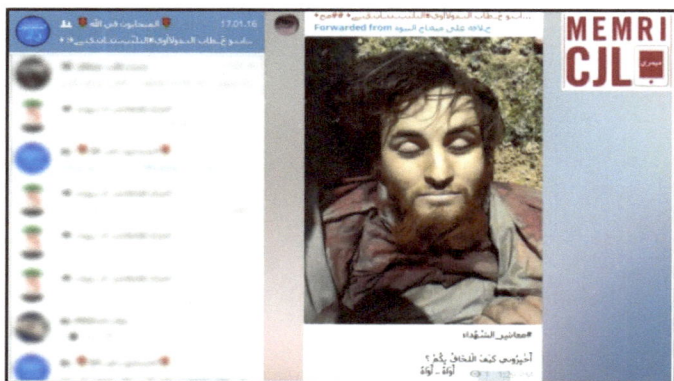

Shared on Telegram on January 17, 2016:: "#Companies_of_Martyrs, tell me how to follow you?..."

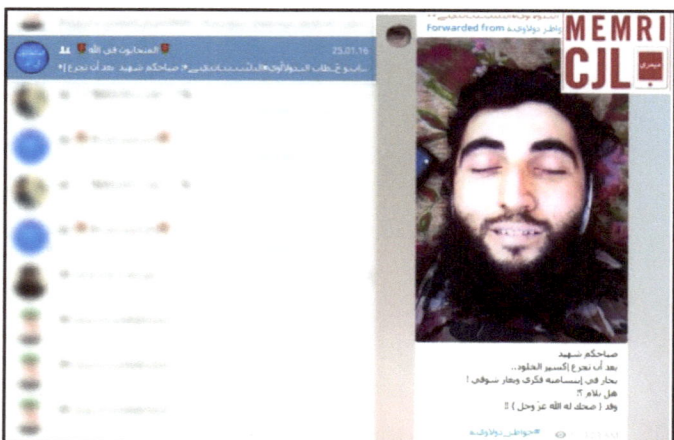

Shared on Telegram on January 25, 2016: "...After he sipped the elixir of eternity, [he shows] oceans in his smile..."

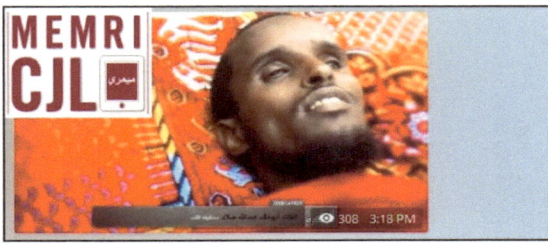

Shared on Telegram on March 5, 2016: "What have you see that you smiled this smile? I wish we'll smile like that when we die. What comfort have you seen (May Allah accept you, O martyr of Islam)"

Shared on Telegram on March 21, 2016: "Abu Shaheed: ...Tell me about the breeze of Paradise, the taste of rivers, and about your seat next to Allah"

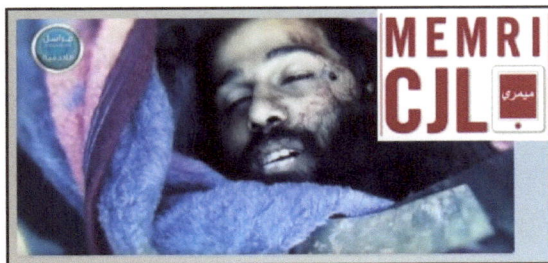

Shared on Telegram on March 25, 2016

Shared on Telegram on March 28, 2016

Shared on Telegram on April 11, 2016: "Abu Shaheed: ...Tell me about the breeze of Paradise, the taste of rivers, and about your seat next to Allah"

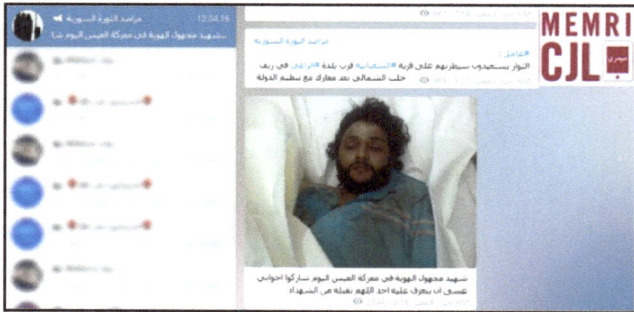

Shared on Telegram on April 12, 2016: "An unknown martyr during the battle of Al-ʿIees today. Participate O brothers for hopefully one will recognize him, May Allah accept him among His martyrs"

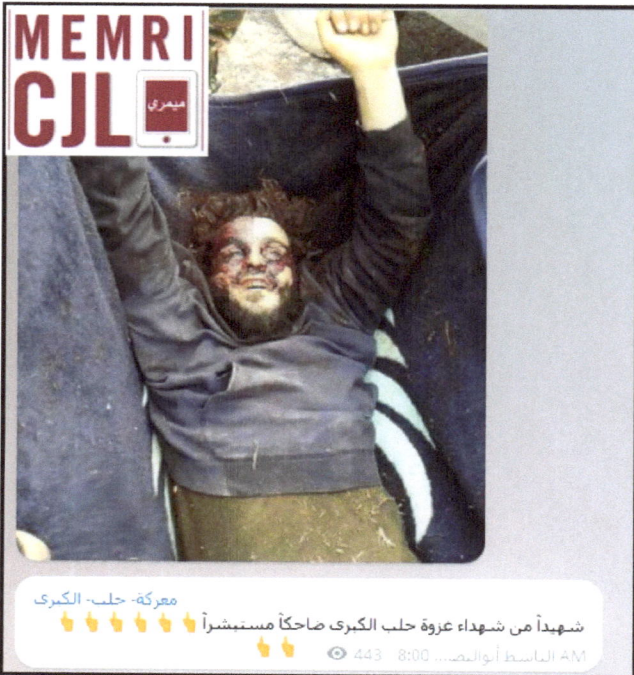

Shared on Telegram on August 5, 2016: "One of the martyrs of the Great Aleppo battle, laughing and seeing glad tidings"

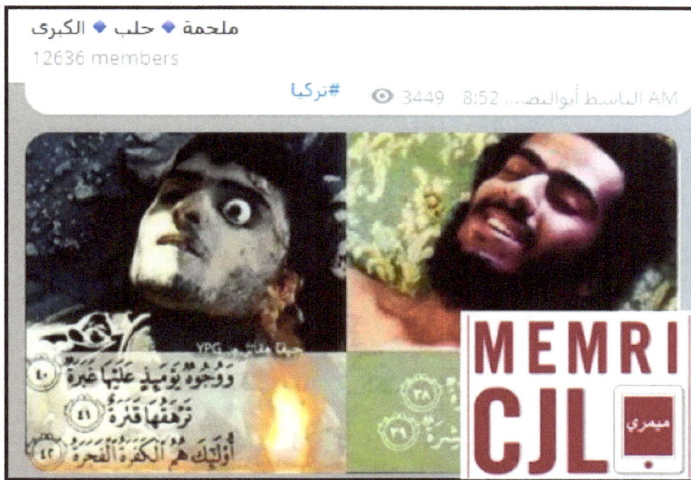

Shared on Telegram on August 22, 2016. Note: The individual on the left is a Kurdish YPG fighter who died fighting ISIS. This image was aimed at illustrating how fighters who martyr themselves for ISIS look peaceful in death, whereas those who challenge the group rot and decay.

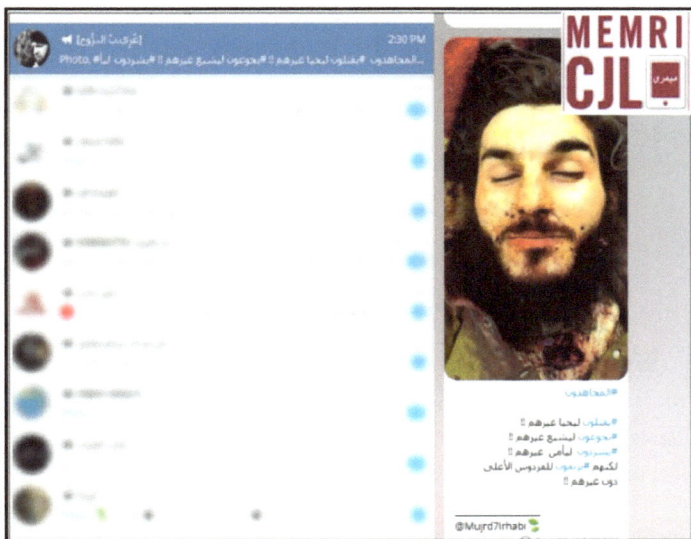

Shared on Telegram on September 30, 2016: "#The_Mujahideen #Killed so that others live, #Starve so that others are full, #Displaced so that others become safe, but they #Ascend to the higher Paradise without others!!"

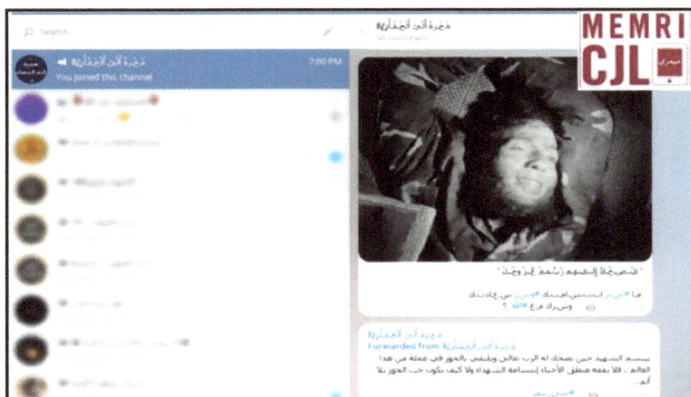

"The Shaheed smiles when Allah the Almighty laughs for him, and he meets the virgins [of Paradise] with disregard to this world. He does not comprehend the logic of the living..."

WARNING – EXTREMELY GRAPHIC: "Strangers" Telegram Account Posts Pictures Of Brutal ISIS Executions For Western Sympathizers, September 8, 2016

A channel on Telegram known as The Strangers specializes in sharing ISIS propaganda and news updates to its sympathizers, whom they call "Al-Muahideen [monotheists] in the West." The channel was created on March 17, 2016, and as of September 1, 2016, it had 60 members, and had posted 8,252 photos, 135 video files, 118 audio files, 2,347 other files, one voice message, and 906 links. It shares updates regarding ISIS campaigns in Iraq and Syria, as well as the news regarding punishments meted out to those within ISIS territory who violate shari'a law. The photos below, posted on April 5, 2016, show ISIS members carrying out beheadings, crucifixions, and shootings.[119]

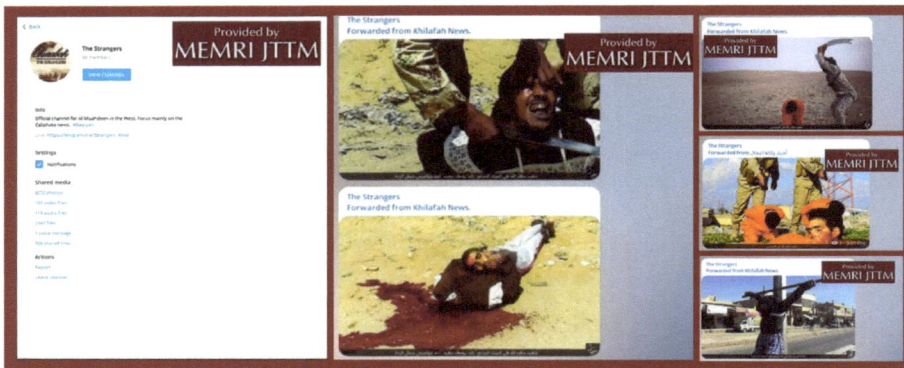

In Wake Of Battle To Liberate Mosul, Pro-ISIS Writer Notes On Telegram Channel That "Hundreds Of Mujahideen Cannot Stop" The Infidels, Condemns Online Jihadis For Staying Home, September 28, 2016

On September 27, 2016, the Arabian Islamic Maghreb Group, a pro-ISIS Telegram channel, published an article criticizing Muslims for not taking part in the fight against the infidels and urging them to join ISIS or carry out operations where they live. The article, which the channel called "very important," also criticized online jihadis and ISIS supporters operating on social media for not making *hijra* or carrying out attacks, and only disseminating content and encouraging others to join ISIS.

Criticizing online jihadis for being too comfortable with merely inciting and encouraging others to wage jihad and join ISIS, and describing them as "those who would encourage people to pray and publish [content about] the punishment for those who abandon prayers but who themselves do not pray," the article added: "I fear that someone will march forth [to wage jihad] because of your encouragement, and will come to Judgment Day as a martyr who will not be judged, while you will be standing there before Allah and asked about your choice to stay behind and abandon the Muslims."

The author also condemned the migrants who had died en route to Italy, lived in forests in Spain, or walked thousands of miles to Germany so that they could live. "In the meantime," he said, "the monotheist has not even bothered to find a way to disturb the mood of the infidels."[120]

ISIS Supporters Create Matrimonial Group On Telegram, July 6, 2016

The "Baqiya Matrimony" group, aimed at pairing up ISIS supporters, was created June 9, 2016 on Telegram, and was also advertised on Facebook. The name was soon changed to the less-conspicuous "Love Fillah." Those seeking a mate must fill out a questionnaire; the questions are provided in the Telegram group. According to the group, this is a safe and secure way to go about finding a spouse, because there are "those who hide as spies online."[121]

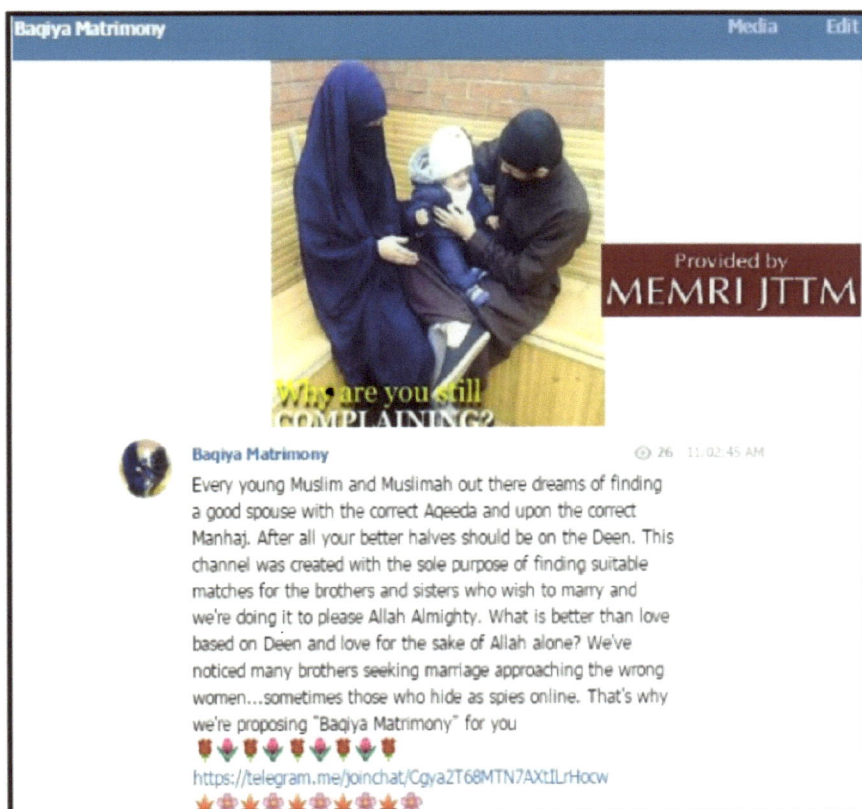

ISIS Fighters, Supporters Following Launch Of Campaign To Retake Mosul: Claims That Enemy Being Pushed Back Due To Dozens Of Martyrdom Attacks; Calls For Lone Wolves To Attack New York, New Jersey, London, Paris, Ankara, Tehran As Revenge, October 18, 2016

Since the start of the military operation to retake the city of Mosul on October 16, 2016, ISIS and its supporters have also been trying to fight back on the informational level, increasingly reporting on its fighters' achievements on the battlefield and on the dozens of martyrdom attacks carried out against enemy forces. This informational activity, mostly on Telegram, is aimed at boosting fighter morale and at disproving claims that the organization is being defeated on the ground. Both official and unofficial ISIS sources have posted infographics detailing ISIS achievements in battle, including the number of martyrdom attacks, enemy losses, and various battlefield stats.

Some ISIS supporters also called on lone wolves to take vengeance on the "crusaders" for participating in the Mosul campaign and carry out revenge attacks in New York, New Jersey, Paris, Ankara, London, and Tehran. One ISIS supporter stressed that Allah required the believers to wage a jihad war against His enemies, but did not promise them victory. This statement could indicate that some ISIS supporters are aware that the chances of thwarting the attack on Mosul are slim, and therefore stress that the path of jihad is more important than victory.

Official communique by ISIS in Ninawa Province claiming that its fighters managed to block an attack east of Mosul, including by using nine car bombs

Alongside official ISIS publication, there have been increased comments on social media in general, and on Telegram in particular, by ISIS supporters. The ISIS-affiliated Al-Nusra Al-Maqdisiyya channel, which appears to operate out of Gaza, published a post stressing that it is not victory that is most important, but rather the path of jihad for the sake of Allah. This seems to be laying the groundwork for an ISIS loss in Mosul, and therefore presents religious arguments that downplay the loss. It stated: "Allah the Almighty guided us to wage jihad war according to His path and fight His enemies. He did not order us to win.... The outcome is in the hands of Allah. Know that we will not win based on the number [of soldiers] or [the size of] our arsenal. We will only win by returning to our faith and our Creator."

Another ISIS supporter called on would-be lone wolf attackers to take vengeance on the enemies that attacked Mosul: "Oh, lone wolf: The Byzantines [the West] and the Persians [the Iranians] attacked Mosul! So rise up and spark their own Mosul in New York and New Jersey! In Paris and London! In Ankara and Tehran! Rise and defend the Islamic State..." Another post on the same channel stated: "The Persians and Byzantines attacked Mosul together! So attack them throughout Mosul! Attack the hordes of Byzantines and Persians wherever they are, especially in the heart of their homes, as this is the worst and most painful thing for them!"[122]

Post on Al-Nusra Al-Maqdisiyya Telegram channel calling on lone wolves to attack the West, Ankara, and Tehran; Al-Nusra Al-Maqdisiyya post calling to attack the "Byzantines and Persians" wherever they are.

Pro-ISIS Telegram Account Offers Tactical Tips To Lone Wolves In Saudi Arabia, October 7, 2016

On October 5, 2016, the pro-ISIS Telegram channel Wahj Al-Jazrawieen published a document featuring tactical tips for lone wolves in Saudi Arabia, for before, during, and after jihadi operations. The document, titled "Tips to the Lions of the Peninsula," provided detailed logistical and operational tips to prospective attackers including the type of materials they need to carry, what food to eat, and how to use ammunition. It stated, inter alia: "Plan well for the operation, survey the location you intended to target for no less than a week and also plan where to go after the execution"; "Carry a backpack that does not impede your movement (such as school bag) to store food and water bottles. The brothers prefer Snickers [the chocolate bar] as one bar equals a meal. It was reported that Sheikh Youssef Al-Ouyairi, may Allah accept him, used to say: I find it wonder-some for a mujahid to not carry Snickers in his bag. And we, oh soldiers of the Caliphate, should never enter a battle without securing at least one or two for each one of us"; "Avoid wearing an outfit in these three colors: Black, red or white, as they easily expose the brother to the enemy and [instead] wear brown or desert-colored clothes, depending on the environment"; "The brother should not throw any trash after him such as paper bags or empty water bottles because they would lead the enemy to his location. He [instead] should conceal them by burying them or placing them inside the backpack and he should even bury his excrement"; and "My dear beloved brother, we know that you have left your house seeking to gain Allah's approval and be martyred for His sake. Think about prolonging the fight with the enemy. The longer you remain steadfast and extend the fight with the enemy, the better. You should extend [the fight with the enemy] before martyrdom."[123]

Tactical tips.

Pro-ISIS Telegram Channel Calls For Lone Wolves to Target Americans In Saudi Arabia, Saudi Intelligence And Security Personnel, October 25, 2016

On October 23, 2016, the Pro-Islamic State Telegram channel Al-The'ab Al-Munfareda ("Lone Wolves") posted a message calling on supporters to carry out attacks on Americans who are in Saudi Arabia, as well as members of Saudi state security and investigation apparatus using guns, poison and knives. The Channel further explains the goals and responsibilities of the lone wolves in Saudi Arabia and other parts of the world, urging them to count on Allah and not to be discouraged if the news about attacks they manage to carry out does not reach the Islamic State and its soldiers. However, the channel assures to its supporters that their actions will be valued by the Caliphate. The channel notes the value of lone wolf operations, saying they are hard to predict, especially since they are carried out by unknown individuals, and that they "disperse the security focus of the tyrants." The channel assures to its members that they can carry out such attacks in Saudi Arabia or any part of the world, while noting that "keenness for obedience to Allah" is among the main reasons for doing so.

The message briefly highlights the lone wolves' responsibilities and suggests different methods to carry out their attacks, including by using kitchen knives, explosives, poison, and guns. It also reminds them not to be discouraged by their lack of knowledge, for example, in manufacturing explosives or booby-traps, and points out that the Internet is full of tutorials on the topics. It says: "Or even if you didn't have anything, don't you have a kitchen at home? Very good. Go [there] and grab the knife from it and head out to the enemy of Allah... and stab him several times in the abdomen, and finish him in the neck."

The channel also asks those planning to undertake such operations not to be saddened if the Islamic State isn't aware of them or their actions, while reassuring them that their actions will be undeniable after the fact by the Islamic State and its soldiers, and that it will earn them their ticket into the group. It says: "In a nutshell, this is your job. Yes, this is it, oh lion of the [Arabian] Peninsula and its hidden wolf. And don't be saddened if the caliph or one of his soldiers didn't know you, for it is enough that Allah knows you, [for] after which, neither the caliph nor his soldiers will cover their ears after hearing [the news] about you claiming the head of an American, Crusader, or members of the [Saudi] state security and investigation [apparatus] in Al-Qassim, Jeddah, or Riyadh."[124]

Encouraging, Promoting, And Praising Lone-Wolf And Other Terror Attacks In The West

"Self-Identified American ISIS Fighter On Telegram Encourages Followers To Carry Out Lone-Wolf Attacks Instead Of Immigrating To Islamic State, June 1, 2016

On May 31, 2016, pro-ISIS American jihadi Telegram user Amriki Muhajer, who claims to be an ISIS fighter, posted on his Telegram account a message encouraging his followers to carry out lone-wolf attacks, as opposed to immigrating to the Islamic State, which fighters have frequently called for doing in the past. Amriki Muhajer notes in his post that he is contacted daily by interested individuals who would like to make hijrah (i.e. immigrate), wage jihad, or kill infidels.[125]

Amriki Muhajer
Provided by MEMRI JTTM

Description
This Channel is dedicated to Al-Muhajreen to the land of Caliphate from the West. Focuses on news of Al-Muhajreen Al-Muwahdeen in Dawlah.

In Video, ISIS Fighters, Including An American, A Frenchman And Russian, Praise Orlando Shooting, Call For More Attacks In West, U.S., Russia, June 20, 2016

On June 19, 2016, ISIS's Al-Furat Province in Iraq released a six-minute video via Telegram titled "You Are Not Held Responsible Except For Yourself," featuring five ISIS fighters from different countries, each of whom delivers a message in his own language. The video is subtitled in Arabic. The main speaker is an American called "Abu Isma'il Al-Amriki"; the others are from Russia, Indonesia, Uzbekistan and France. The basic message of the video is that ISIS does not fight only in the areas it controls, but has brought the battle to the West, to Russia and to other countries. The speakers in the film praise Orlando shooter Omar Mateen, calling him "a soldier of the Caliphate." They direct threats at the U.S., Russia, France and other "Crusader" and "infidel" countries, and call on Muslims in these countries to carry out attacks there. The last part of the film is a segment from the video made by Abballa Larossi, who murdered a police officer and his wife in France, in which he too called for attacks. The following are excerpts from the video and further details about it:

ABU ISMA'IL AL-AMRIKI
Provided by MEMRI JTTM

Abu Isma'il Al-Amriki

The video begins with excerpts from news reports on the Orlando shooting. One, from Al-Jazeera, mentions that a video released by ISIS's Al-Furat Province in the wake of the November 2015 Paris attacks had threatened that one of the organization's next targets would by the U.S. The first speaker in the film, the American called "Abu Ismail Al-Amriki," promises the U.S. that despite its power it will eventually be defeated by the Islamic State. Speaking in English, he says: "Oh America, do you think you are at war with a small group of mujahidin in Iraq, Syria, Libya and other places? You are sadly mistaken... Oh America, indeed you are at war, with all the true and sincere Muslims around the world who yearn and desire to see the honor of Islam return... with the people who wish to be killed and slain for the sake of Allah... [and] with the people who have had this creed in their hearts since the Prophet of mercy and war, Muhammad, [was] sent to this world, and we will remain until your forces will be crushed, defeated and humiliated in Dabiq, with Allah's help."[126]

Following Orlando Shooting, ISIS Supporters Gloat, Threaten Further Attacks On U.S., June 13, 2016

Shortly after the June 12, 2016 shootings at the Pulse nightclub in Orlando, FL, and especially after A'maq announced that the shooter, Omar Mateen, was an ISIS member, ISIS supporters on Telegram began posting messages that included gloating, praise for the shooter, and banners and images promising further attacks on America and its symbols. Some supporters said that another attack was imminent in the U.S., and that it would be surprising in quality. The supporters all highlighted that this attack took place during Ramadan, which is considered the month of jihad. The banners and images posted by ISIS supporters included direct threats to target the White House, Washington DC, and even California. Some comments included justifications for killing Western civilians, echoing statements made by ISIS spokesman Abu Muhammad Al-Adnani in a recently-released audio recording. Other comments also included attacks on the LGBT community, and criticized those in the Muslim world who expressed solidarity with it. Concurrently, Al-Qaeda supporters also expressed joy at the shooting, calling for more attacks in America.[127]

Following Terror Attack In Nice, ISIS Supporters Post Banners Depicting Threats To France, U.S. Capitol, And Gloating Over France Tragedy, July 19, 2016

After the July 15, 2016 Nice, France truck attack, ISIS supporters posted dozens of banners on various Telegram channels gloating over it. The banners threaten that ISIS will continue striking France until it conquers the country, raises its flag over the Eiffel Tower and on the roofs of Paris' most notable landmarks. One banner mockingly states that up to now the French were frightened by the sight of a bearded Muslim or a person dressed in Muslim attire, but that now they are hallucinating about trucks. Below are some examples:[128]

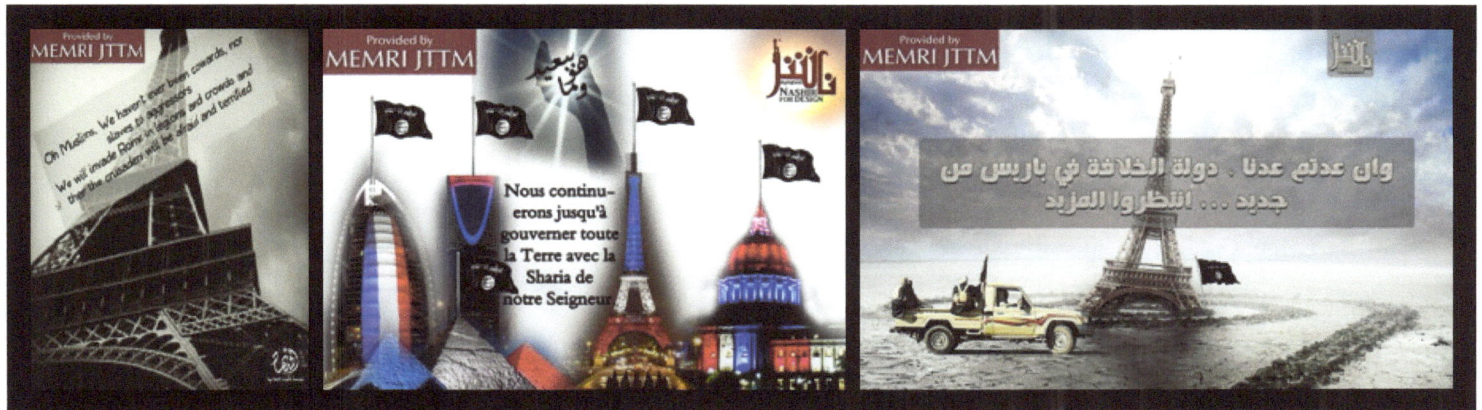

Video By Media Company Associated With ISIS Threatens Attacks On Western Cities – Including New York, Paris, Rome, Berlin, Moscow; "We Have Lone Lions That Stalk You," July 14, 2016

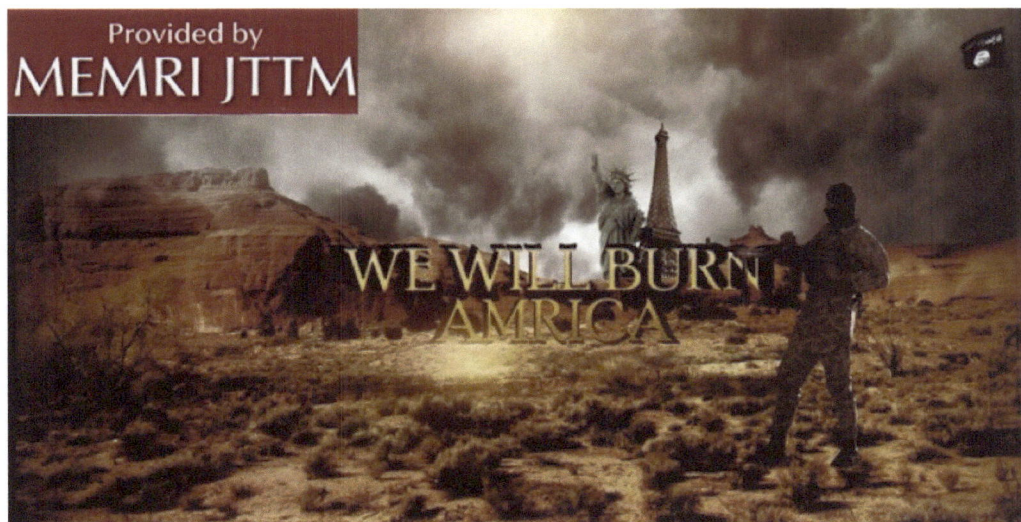

On July 13, 2016, Al-Waa'd Media Production, a media company associated with ISIS, published a seven-minute video titled "In Your Own Homes." The video, which was also posted to the Al-Khilafa Telegram channel, includes threats of attacks on central Western cities. The video is also accompanied by nasheeds (Islamic songs) in French that were recently published by ISIS's Al-Hayat Media Center, praising the group's attacks in Paris and Brussels and promising further attacks in the future. The video shows the casualties and destruction caused by coalition airstrikes, interspersed with scenes from ISIS beheading videos as well as footage from the recent ISIS attacks in Europe, which it claims were retaliations for Western aggression against ISIS and Muslims in general. The video also includes simulated scenes accompanied by captions such as "the fight just began," "we will burn America," and more. One caption poses the question of which city will be targeted next, suggesting it could be Rome, Berlin, or maybe Moscow. Staged scenes of fighters selecting weapons and preparing bombs are shown to illustrate that ISIS is preparing to attack Western cities. The video states: "Oh, nation of the cross, this is a message for you: You launched aggression and attacks against the mujahideen and peaceful [Muslims]. We will strike terror and fear in your own homes. Blood for blood and destruction for destruction. We will fight you in your own homes. We will not remain silent for the fact that you have caused chaos in Muslim lands. We have lone lions that stalk you. The solution for you is simple: Either leave us be and convert to Islam, or pay the jizya poll tax. This is the law set by our Lord. And now that the war has not been won, and is only just beginning."[129]

French, Bosnian ISIS Fighters Urge Muslims In U.S., Canada, Australia, Europe To Carry Out Attacks, June 27, 2016

On June 26, 2016, ISIS's Ninawa Province released a nine-minute video via ISIS's official Telegram channel titled "Most Beneficial to Us and Most Harmful to Them," celebrating the Orlando attack and the murder of a French police officer and his partner in a suburb of Paris. The video features two French ISIS operatives and one Bosnian, who praise the Orlando and Magnanville attackers and call upon Muslims in the West to follow their example. One of the French-speaking fighters also rails against homosexuals in justifying the Orlando attack. The other addresses the citizens of the U.S. and France, urging them to take to the streets to demand an end to their countries' attacks on ISIS, just as the Spanish people demanded to end their government's involvement in the Iraq war after the 2004 Madrid bombing. The video is narrated in Arabic; the fighters speak in their native languages and are subtitled in Arabic. The following are excerpts:

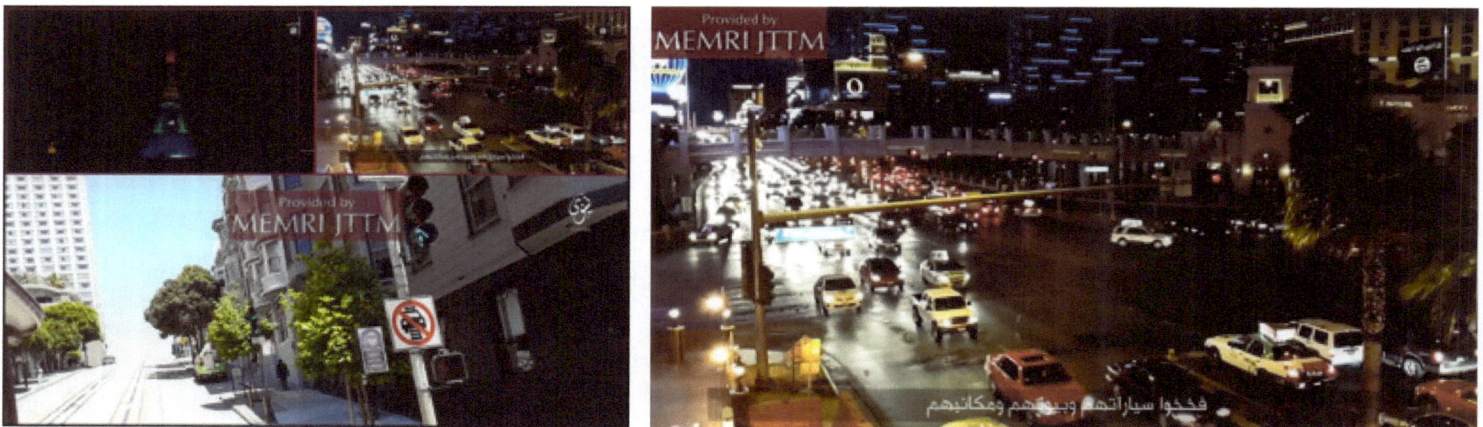

Footage from San Francisco and elsewhere in the U.S.; footage of Las Vegas, with the caption "Booby trap their cars, homes and offices"

The first speaker, Abu 'Abdallah Al-Bosni ("the Bosnian"), calls upon Muslims in the West to carry out attacks: "Oh servants of Allah, oh believers in America, Australia, Canada, Europe and especially the Balkans, and in all the lands of the infidels! Do you not remember the hadith of the Prophet, who said: *The believers* in their mutual kindness, compassion and sympathy are like a single *body*. When one limb suffers, the whole body responds with wakefulness and fever.' Oh Muslims in the West, the time has come for you to awaken from your dream. The time has come for you to awaken and rebel... Rise up, oh Muslims, and kill the Christians and their agents in their own lands. Booby trap their cars, their homes and offices. Kill them with sniper rifles and silencer guns. Oh Muslims, kill them however you can, even if with a knife. Do not distinguish between those who are soldiers and those who are not... because their jets do not distinguish between combatants and non-combatants in the Islamic State..."

Al-Bosni's message is accompanied by footage of San Francisco, made to seem like reconnaissance footage filmed for the purpose of planning an attack. He continues:

"I would like to commend my brothers that I love in Allah. [First] my brother 'Omar Mateen, who killed in Orlando more than 50 people who engaged in the act of the people of Lot [i.e, homosexual relations]. O my brother, you gladdened our hearts, you gladdened the hearts of Muslims, while sparking hate in the [hearts of the] hypocrites in the Muslim community... Allah be praised, I also wish to pay homage to the brother Abdullah Laroussi [the killer of the policeman in Paris]. May Allah accept you, my brother. You were a real man, you were a man, you answered the call of Allah, you gladdened my heart and the hearts of Muslims..."[130]

Poem By Female Jihadi Praises Orlando Attack, Threatens U.S., July 7, 2016

On July 6, 2016, the pro-ISIS Telegram channel of Al-Sumoud media company published a poem praising the June 12 Orlando attack. The poem appeared on a poster showing the White House in crosshairs with the title "Raid On America," a reference to the attack on the Pulse nightclub in Orlando, FL. The poem was written by Ahlam Al-Nasr, allegedly a female jihadi.[131]

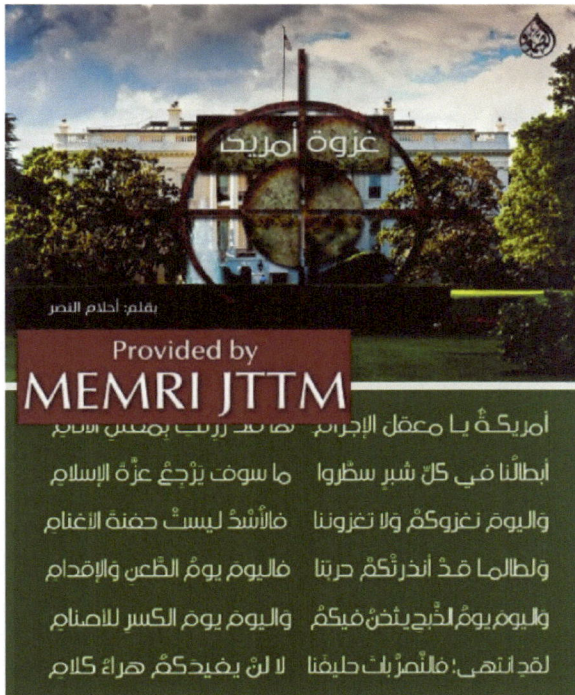

Pro-ISIS Libyan Media Activists Post Images Depicting Threats To Paris, Rome, May 26, 2016

On May 24, 2016, the pro-ISIS Libyan media group Libya Waz Al-Khilafah published on Telegram a series of graphic images containing threats to Paris and Rome. The group also previously published content in support of ISIS's Libya branch. The images feature quotes in Arabic from the most recent speech by ISIS spokesman Abu Muhammad Al-Adnani, with the Eiffel Tower in Paris and the Colosseum in Rome in the background. In the quoted statements, Al-Adnani calls for the mobilization of ISIS supporters in the West to carry out terror attacks. Previously, on May 11, 2016, the same group published a graphic image featuring the late ISIS leader in Libya Abu Mughirah Al-Qahtani, an ISIS motorcade, and a man standing with an ISIS flag atop of the Coliseum in Rome. The caption promises that ISIS will walk in the footstep of its leaders and conquer Rome.[132]

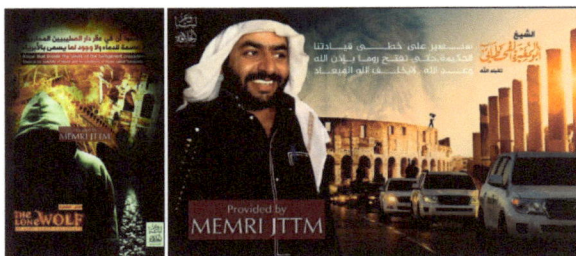

Pro-ISIS Group Threatens Citizens Of Spanish-Speaking Countries: We Will Kill You Wherever You Are Unless You Stop Fighting Muslims, May 30, 2016

On May 30, 2016 the Al-Wafa Foundation, a pro-ISIS activist group on Telegram, posted "A Message to the People of Spain and to the Spanish-Speaking Countries," threatening these countries and their citizens with terror attacks unless they stop fighting the Muslims and occupying Muslim land. The two-page message, dated March 28, 2016 and signed by "Abu Bara bin Malik – servant of the Caliphate," states that ISIS "insists on killing infidels" because their countries wage war on the Muslims and kill innocents, and states "We will kill any innocent 'Spanish infidel' on the spot if we see one of you in a Muslim country" and "[we will] kill you in your [own] cities and villages." The threatening message, which was distributed in Spanish, Arabic, English and French, may be a reaction to a series of arrests of ISIS elements in Spain which occurred in February and March 2016.[133]

Hours Before Euro 2016 Soccer Finals In Paris, ISIS French Media Share Threatening Posters Online, July 10, 2016

On July 10, 2016, French speaking ISIS media supporters distributed images on Telegram threatening the Euro 2016 finals which are due to take place under high security later today in the Stade de France near Paris. One image shows crosshairs on the head of a French soldier patrolling the streets, and the caption is a quote from a speech by ISIS spokesman Muhammad Al-Adnani, calling upon ISIS supporters to carry out attacks in the West: "Know that each one of us would like to be in your place in order to severely punish the crusaders, night and day, without sleeping. In order to frighten and terrorize them until each one is scared of his neighbor."[134]

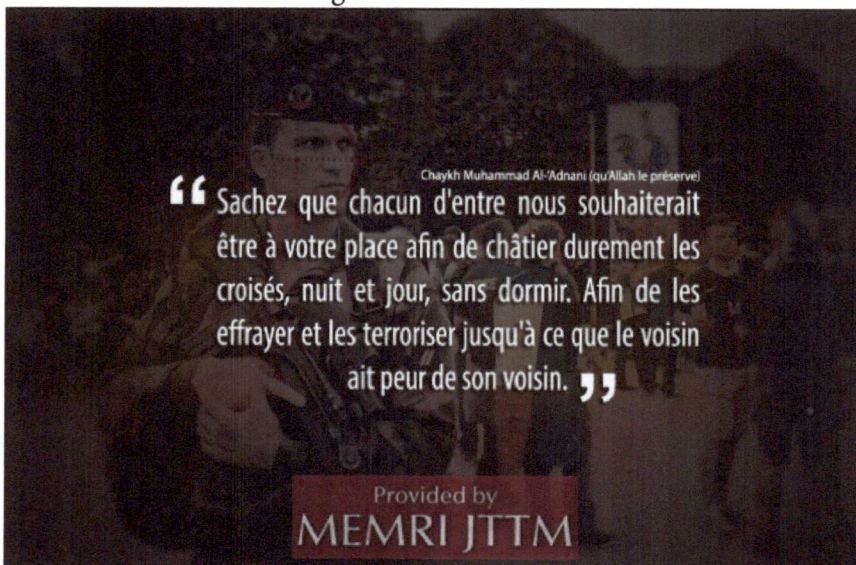

ISIS Supporters Express Joy Over Nice Truck Attack, Claim It Was Retaliation For French, Coalition Airstrikes In Iraq, Syria, July 15, 2016

Following the truck attack in Nice, France on the night of July 14, 2016, which claimed the lives of at least 84 people, ISIS supporters online expressed joy, congratulated the perpetrators, and stressed that the attack came in retaliation for airstrikes by coalition jets, including ones from France, in Syria and Iraq, which killed innocent Muslims. ISIS sympathizers on Telegram also posted banners lauding the attack and vowing that members of the Islamic State would continue to strike France and the West.

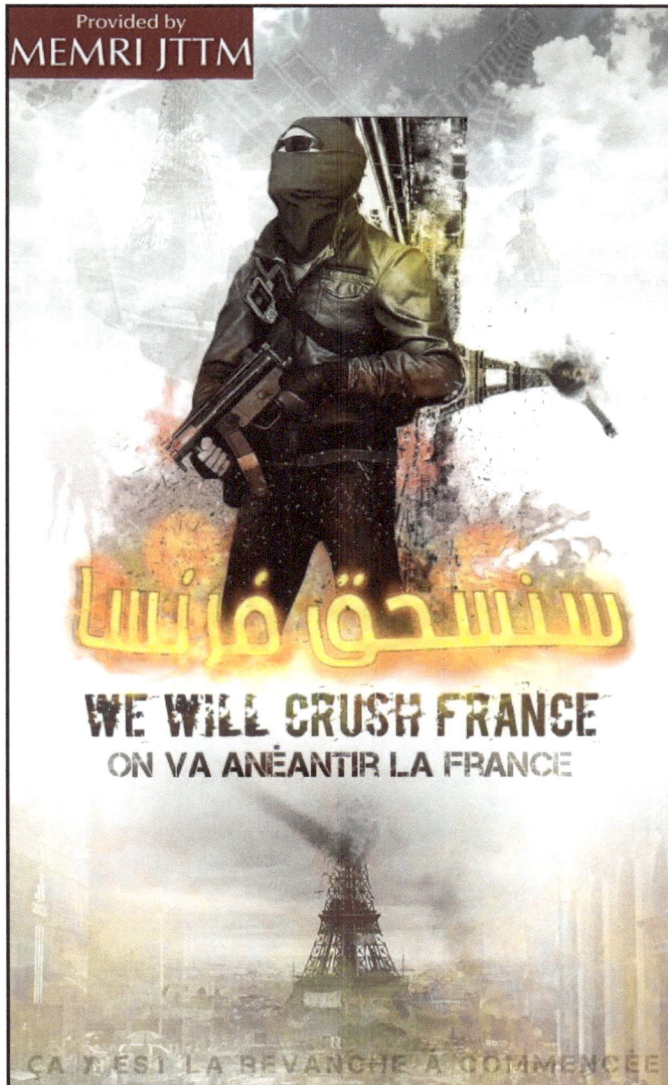

The Dabiq Telegram channel posted a comment by an ISIS supporter with the username miqdad, who wrote: "Whether the Islamic State claimed responsibility for the first operation in Paris or not – and it eventually did – and even if it has yet to claim responsibility for the Nice operation, which is still vague at this point [meaning no one had claimed responsibility for it] – our hearts are still filled with joy of a level we have yet to experience! How can we not rejoice when we see the organs of the infidel Crusaders, who launched a war on Islam, strewn about here and there?! How can we not rejoice after being sad when they bombed Muslim lands in Iraq, Syria, and elsewhere in cold blood?! Do we not deserve joy?!" Pro-ISIS Telegram channels also used the attack as an opportunity to spread messages aimed at intimidating Western citizens.

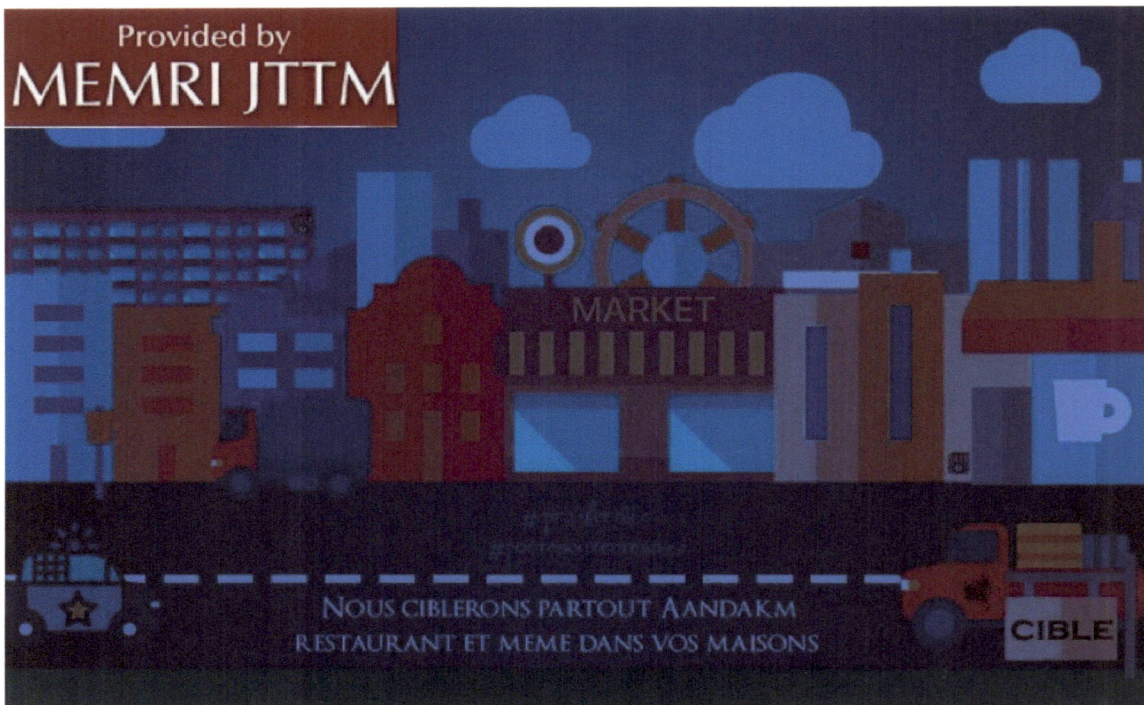

Banner threatening attacks on France

One user posted a graphic of an ISIS fighter strangling French President Francois Hollande as his friends watch and laugh.

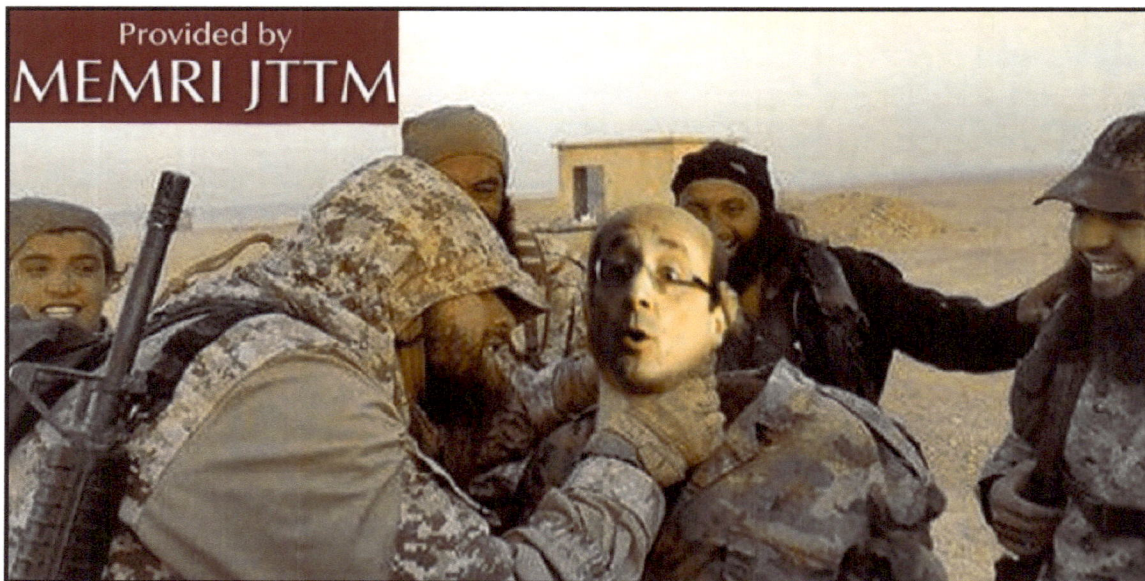

The Kawasir Al-Nasher channel posted a banner in French and Arabic featuring threats to France, which states: "Oh Frenchmen, just as you kill us, so you will be killed. We have promised you that we will move the battlefield to the heart of your home. Then you will taste our wrath." Another banner stated: "Allah protects His believers, Oh Hollande... And here you are humiliated by His abilities and the abilities of His soldiers. In spite of you, we will humiliate you, oh pig of France."

ISIS Operatives Distribute Kill List, Target List, Operational Tips For "Lone Lions" In France And The West, August 15, 2016

On August 11, 2016, Francophone operatives active on Telegram distributed two infographic images with names of individuals and organizations, as well as specific locations, for targeting by ISIS militants in France. These images are part of ISIS's ongoing efforts to capitalize on the recent wave of terror attacks in France and the West and to inspire more supporters to carry out attacks. The repeated threats include, for instance, members of the national French education system such as teachers; journalists and researchers; soldiers and police; Muslim leaders considered to be apostates; musicians; Jewish intellectuals; and so on. The first image, titled "Targeted Attacks," sets out the specific individuals and organizations to be targeted by ISIS operatives. It is divided into sections – journalists, politicians, and the like – and includes the names of the names of the individuals to be killed. The second image, titled "Mass Attacks," offers a target list of public places, ideas for modus operandi, and other information useful to operatives in carrying out indiscriminate terror attacks against the public. The images were originally released by an ISIS operative on Telegram called "Sabre de lumière" [Sabre of Light] and shared by other ISIS operatives on Telegram.[135]

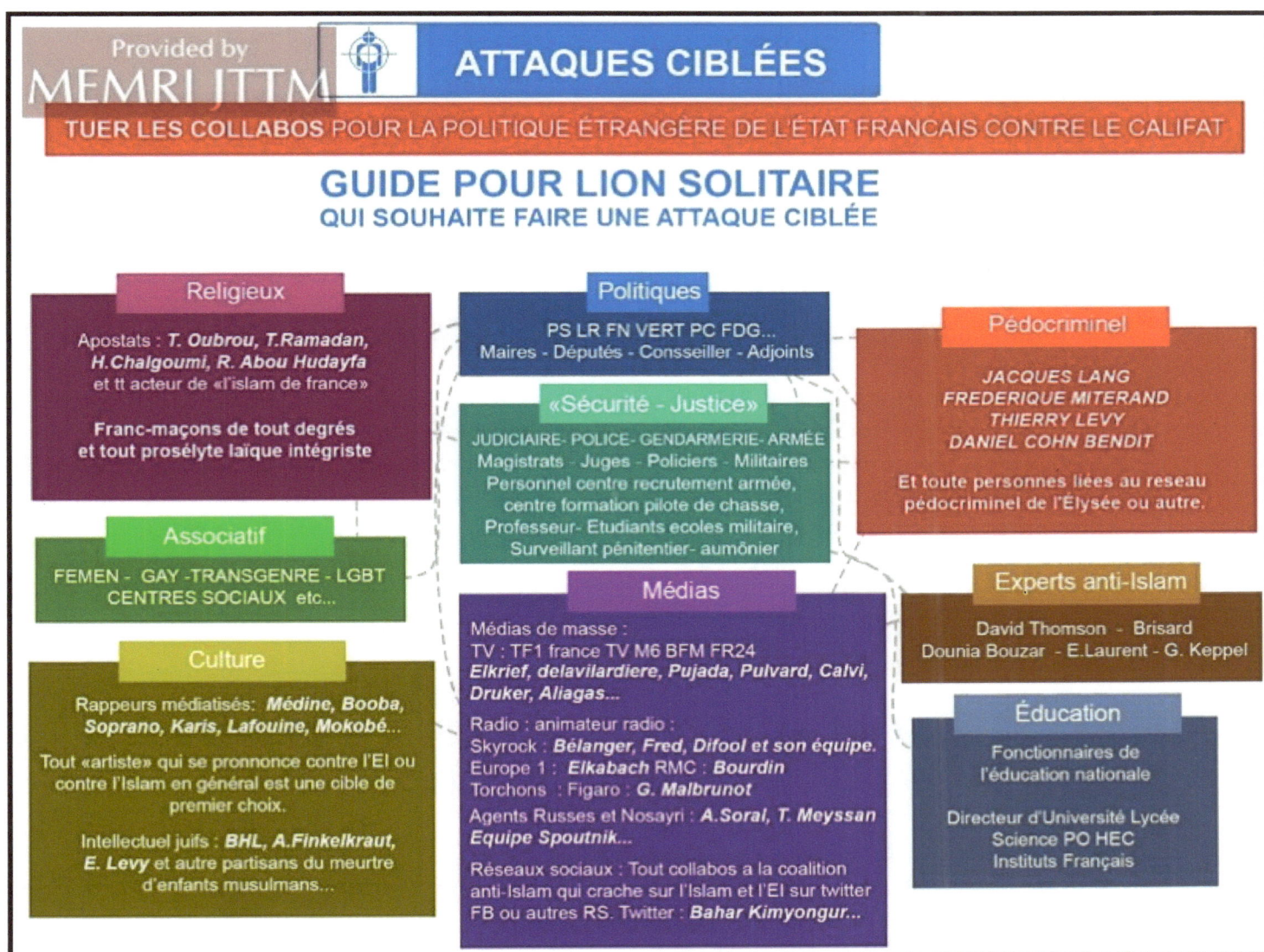

Provided by MEMRI JTTM

ATTAQUES CIBLÉES

TUER LES COLLABOS POUR LA POLITIQUE ÉTRANGÈRE DE L'ÉTAT FRANCAIS CONTRE LE CALIFAT

GUIDE POUR LION SOLITAIRE
QUI SOUHAITE FAIRE UNE ATTAQUE CIBLÉE

Religieux

Apostats : *T. Oubrou, T.Ramadan, H.Chalgoumi, R. Abou Hudayfa* et tt acteur de «l'islam de france»

Franc-maçons de tout degrés et tout prosélyte laïque intégriste

Associatif

FEMEN - GAY -TRANSGENRE - LGBT CENTRES SOCIAUX etc...

Culture

Rappeurs médiatisés: *Médine, Booba, Soprano, Karis, Lafouine, Mokobé...*

Tout «artiste» qui se prononce contre l'EI ou contre l'Islam en général est une cible de premier choix.

Intellectuel juifs : *BHL, A.Finkelkraut, E. Levy* et autre partisans du meurtre d'enfants musulmans...

Politiques

PS LR FN VERT PC FDG...
Maires - Députés - Conseiller - Adjoints

«Sécurité - Justice»

JUDICIAIRE- POLICE- GENDARMERIE- ARMÉE Magistrats - Juges - Policiers - Militaires Personnel centre recrutement armée, centre formation pilote de chasse, Professeur- Etudiants ecoles militaire, Surveillant pénitentier- aumônier

Médias

Médias de masse :
TV : TF1 france TV M6 BFM FR24 *Elkrief, delavilardiere, Pujada, Pulvard, Calvi, Druker, Aliagas...*

Radio : animateur radio :
Skyrock : *Bélanger, Fred, Difool et son équipe.* Europe 1 : *Elkabach* RMC : *Bourdin* Torchons : Figaro : *G. Malbrunot*

Agents Russes et Nosayri : *A.Soral, T. Meyssan Equipe Spoutnik...*

Réseaux sociaux : Tout collabos a la coalition anti-Islam qui crache sur l'Islam et l'EI sur twitter FB ou autres RS. Twitter : *Bahar Kimyongur...*

Pédocriminel

JACQUES LANG FREDERIQUE MITERAND THIERRY LEVY DANIEL COHN BENDIT

Et toute personnes liées au reseau pédocriminel de l'Élysée ou autre.

Experts anti-Islam

David Thomson - Brisard Dounia Bouzar - E.Laurent - G. Keppel

Éducation

Fonctionnaires de l'éducation nationale

Directeur d'Université Lycée Science PO HEC Instituts Français

Advice For "Lone Wolves" On ISIS-Affiliated Telegram Channel: Poison The "Infidels" Food, Air And Water, Spread Panic By Posting False Alarms, August 21, 2016

A message posted August 21, 2016 on Dabiq, a pro-ISIS Telegram channel, includes a series of suggestions for "lone wolves" on targeting "infidels," including by poisoning them, causing road accidents and delivering false warnings to create panic. The author, an ISIS supporter, urges the readers to circulate the message as widely as possible and to add suggestions of their own for killing infidels. The suggestions included: "1. Insert poison into unwrapped foods such as fruits and vegetables in markets frequented by infidels, or poison foods and beverages that are forbidden according to Islam. 2. Pour oil on steep and winding mountain [roads] in foggy conditions, or reverse [traffic] signs near [sharp] curves in mountain roads."[136]

Pro-ISIS Group Urges "Lone Wolf" Operatives In Europe, Especially In France, To Carry Out Attacks Before They Are Arrested, September 4, 2016

On September 4, 2016, the pro-ISIS Nasher Media Foundation posted a message warning ISIS supporters in Europe to take precautions in light of the recent arrests of "lone wolf" operatives in France. It urged them to carry out attacks as soon as possible, presumably before they are arrested as well. The message, released in Arabic, English and French and circulated via Telegram, read: "We urge our brothers, the lone wolves in Europe and especially in France, to beware and take precautions. We received information today about brothers who were arrested before they could carry out their operations. Therefore, we advise you to erase anything related to the [Islamic] State [from your devices], including photos, videos and programs, and rush to carry out your operations before time runs out. Cling to Allah and put your trust in Him, He is sufficient for you and the best disposer of affairs."[137]

Pro-ISIS Telegram Channel Calls on ISIS Supporters Everywhere To Use Cars, Knives, Rocks To Kill Infidels In The West, October 31, 2016

On October 30, 2016, the pro-ISIS Telegram channel "Dabiq" published a post urging ISIS supporters and soldiers everywhere to emulate terrorist attacks that took place in different parts of the world including cities in the U.S, Europe, Asia and Africa, and to use cars, knives and rocks to kill the infidels. The post, also shared on multiple pro-ISIS Telegram channels, urged ISIS supporters and soldiers to avenge the killing of Muslims and jihadi leaders in airstrikes and focus on carrying out operations and not to overthink the outcome. After reminding ISIS supporters of the joy they have felt after hearing the news about attacks in Russia, Germany, Sweden, France, Kenya, Indonesia, Belgium, the U.S. Denmark, Australia, Bangladesh, Canada, Tunisia, Egypt and Saudi Arabia, the post states: "Would it be difficult for you to emulate what Ahmad, Muhammad, Abdallah and Abu had done?... Ask yourself and think about what I will tell you. Look around you, how many infidels can you reach and kill with your car, your knife, or a rock? They are in their country and can be found everywhere." To further encourage supporters to carry out attacks, the post called on them to remember how "they smile and [express] joy when their countries pour their utmost anger into striking Muslims and killing them... Will you be content with them living in joy while they complicate the living condition of Muslims with their crimes and throw parties and gatherings to celebrate the killing of one of the mujahideen's leaders?" The post ends by urging ISIS supporters to be courageous and kill infidels to obtain the glad tidings promised by the Prophet Muhammad, who said: "An infidel would never meet with his killer in hellfire."[138]

Technical Tips And Information By And For ISIS Fighters And Sympathizers

Pro-ISIS Tech Channel On Telegram Posts Video Tutorial Showing How To Get Facebook Pages Suspended By Reporting Them As Belonging To Underage Children, October 20, 2016

On October 19, 2016, Teqani Al-Khilafa, a pro-ISIS Telegram channel dedicated to tech issues, posted a video tutorial providing step-by-step instruction on getting Facebook pages suspended by flagging them and reporting that they belong to underage children. According to the video, made by "King of Android," who goes by @ABOHSEN2016 on the channel, Facebook users can get any page suspended in this way. After providing a Facebook link to a page where users can "Report an Underage Child," the video demonstrated how to fill in the report form, including the owner's name, page's URL, and actual age of the "child" being reported.

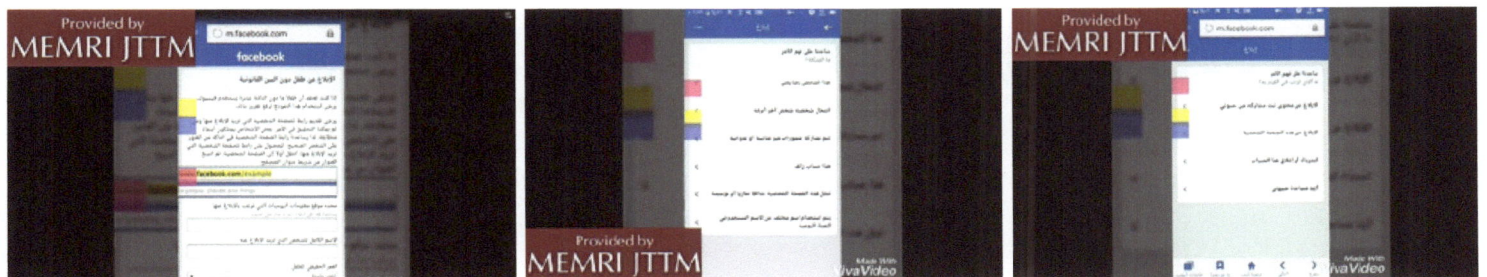

As a jihadi song describing the bravery of the mujahideen plays in the background, the video instructs viewers to paste into the "other" section of the form a note to Facebook founder Mark Zuckerberg, apparently inaccurately translated from Arabic to English, reading "Because it has to penetrate the legal age for these children of Facebook, Mark, I cannot stay in the Facebook."

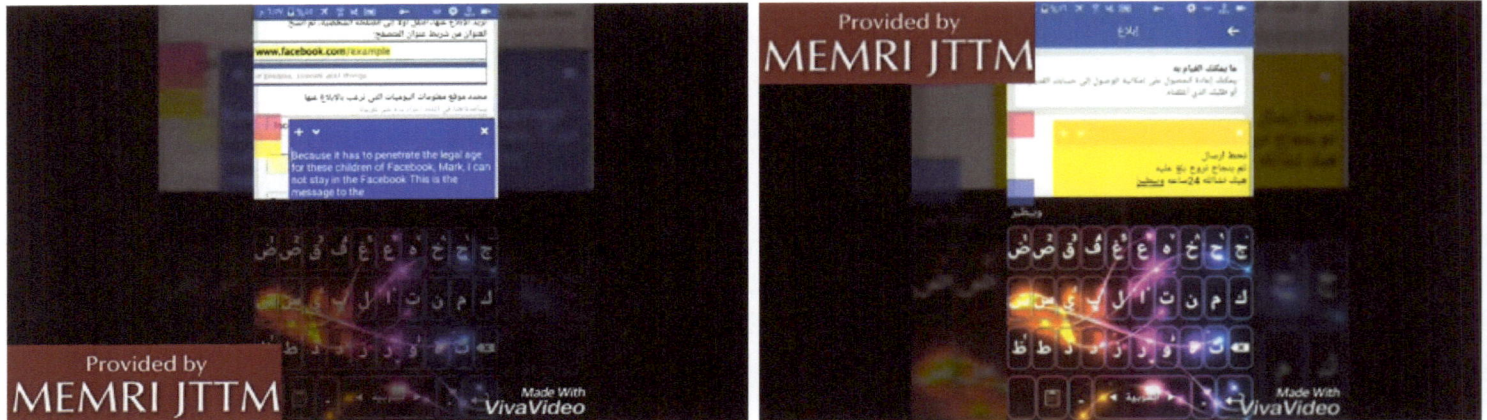

The viewer is instructed to hit "send" and then to go to the profile page of the Facebook page they just reported, click on "more," select "report" from the dropdown list, click on "Report This Profile," then "This Is A Fake Account," and then "block" the user. Then it instructs viewers to return to the page being reported, select "report" again and this time select "Recover or Delete This Account" and then click on "Delete This Account." The video ends with a statement which reads: "Allah's willing, within 24 hours [the account] will be deleted."[139]

Jihadi "Help Desk," Tech Channels On Telegram Offer Tech Support, Tutorials, Up-To-Date Cyber Security Info, February 8, 2016

Jihadis are enjoying the services of a technical "help desk" on Telegram. The help desk is part of a larger network that caters to jihadis' technical needs, and includes channels and accounts on those platforms and on a forum dedicated to technical matters as well. In the past year jihadis have intensified their effort to provide their counterparts with technical know-how on a variety of topics such as mobile phone security, and cyber security-related information in general. The information has generally focused on improving jihadis' cyber security knowledge and awareness with regard to their operations online. Most recently, these efforts have culminated in the establishment of the Electronic Horizon Foundation (EHF), a joint effort of several entities like Tiqani Al-Dawla Al-Islamiyya (the "Islamic State Technician"), a top disseminator of cyber security information, and and the Information Security channel, a technical channel on Telegram. One of the EHF's top goals is to impede electronic surveillance of the mujahideen by Western intelligence services. The nexus operates primarily on Telegram but has a presence on Twitter as well.[140]

Thousands Of ISIS Accounts On Telegram Warn: Don't Make Plans For Migration To Caliphate On Encrypted Messaging Platforms, July 22, 2016

On July 20, 2016, multiple pro-ISIS accounts on the messaging app Telegram shared a graphic image warning about "Hijrah to Khilafa from Darul Kuffar," telling followers to refrain from seeking ways to reach the Islamic State via Telegram. The Telegram accounts also urged their followers to re-share the graphic. The message on the graphic read: "Warning – There is no Hijrah through Telegram or Whatsapp. Whoever claims to offer hijrah facilitation services is a spy wanting to trap brothers. Please Spread."[141]

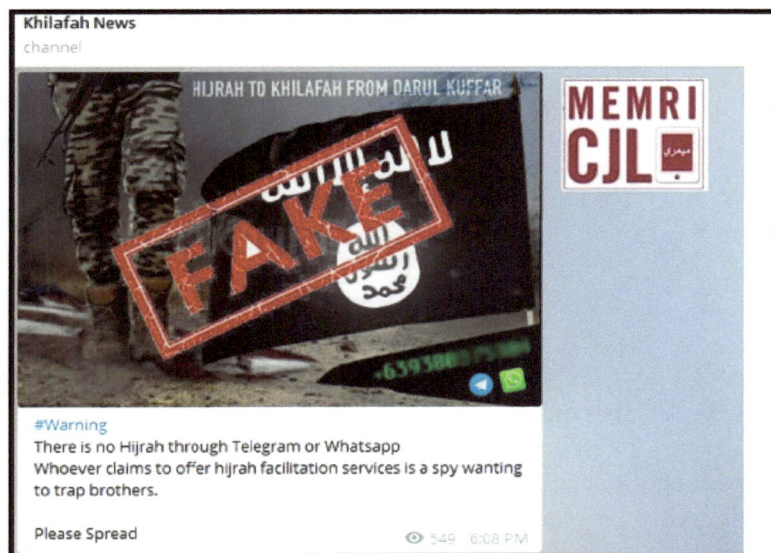

American Jihadi Shares Safety And Encryption Tips On Telegram, May 4, 2016

On May 3, 2016, Amriki Muhajer, an American pro-ISIS jihadi Telegram user who appears to be a member of the hacking group Caliphate Electronic Army, shared a post including safety tips for Telegram. The tips in the message include:

– Registering with a fake number. The post suggests apps such as Talkatone and NextPlus, which can generate fake numbers. It also suggests changing the number as frequently as possible.

– Turning the device encryption on in the settings menu and adding password protection to Telegram, including 2-step verification.

– Using TOP or other VPN to hide the location.[142]

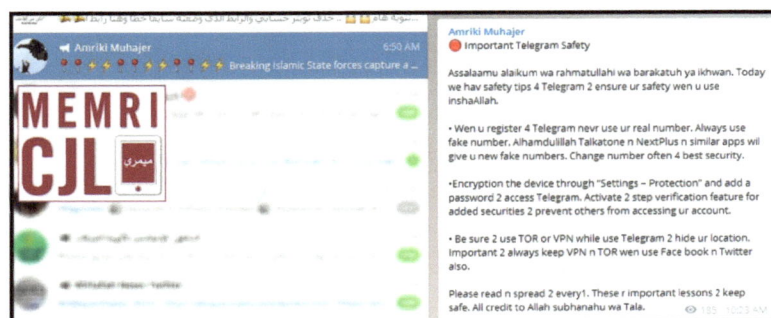

Pro-ISIS Telegram Channel Advertises For Volunteers To Translate, Write, Design, August 5, 2016

On August 4, 2016, the Telegram channel Caliphate News posted a message stating that it was in dire need of volunteers, both men and women, to help with translating, writing, and designing content. The channel provided a username for volunteers to contact (redacted by MEMRI).[143]

ISIS Releases Android App Teaching Supplications To Children, June 23, 2016

After releasing an app last month teaching children the Arabic alphabet, ISIS has released another app, on pro-ISIS channels on Telegram, targeting children, this time teaching them supplications. The app heavily incorporates interactive ISIS and jihadi themes, allowing children, for example, to blow up enemy fighter jets and American tanks. The app works on Android devices, and was developed by ISIS's publishing house, Al-Himma Library. It promises "engaging education" for children through the learning of "over 40 supplications." A desktop version of the app was also released. The app features 42 supplications, arranged in separate icons. The supplications cover a wide-range of situations during which Muslims recite an appropriate prayer, such as when eating, entering and exiting a mosque, visiting the sick, burying the dead, entering a market, going on trip, and many more.[144]

Banner promoting the new app

Pro-ISIS Engineers, Scientists Collaborate On Projects In Telegram Channel, March 18, 2016

On February 20, 2016 a Telegram channel titled "Islamic State Scientists & Engineers" was launched. The channel administrator stressed that members of the channel must have a BSc in a scientific or mathematics field such as chemistry or aeronautics. He stated that the channel has numerous goals such as to "exploit some channel members' situation to do research for the military benefit of the caliphate," and "collect as much caliphate scientists & engineers as possible from around the world & introduce them to each other." The first message in the Telegram channel states, "U can join this channel or group only if u have pledged [allegiance] to the caliphate & have at least a BSc in one of the following fields: Electrical engineering, Mechanical engineering, Chemical engineering, Aeronautics, Physics, Chemistry, Biology, or any closely related subject." One post defined the goals of the channel: "1- collect as much caliphate scientist & engineers as possible from around the world & introduce them to each other. 2- use them to create a powerful worldwide industrial network to support the military industry in the Islamic State. 3- support the scientific education in the caliphate. 4- exploit some channel members situation to do research for the military benefit of the caliphate." Two members were identified by the channel as an aeronautics engineer and as a physicist..

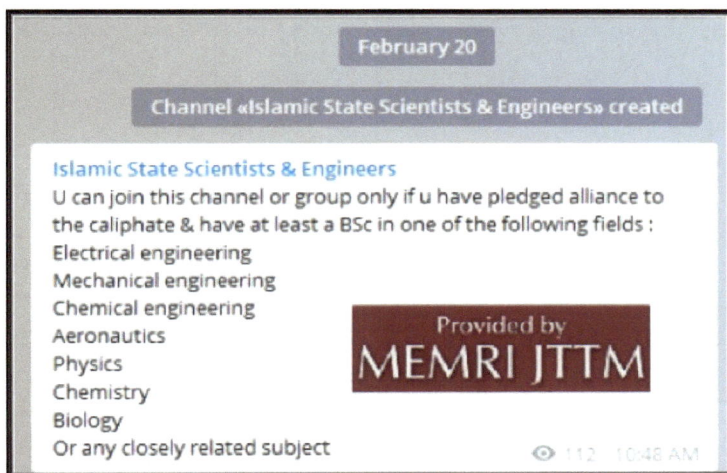

On February 25, the administrator posted the details of some upcoming projects for the members to collaborate on, including bullet manufacture, small turbofan engines, "klystrons & magnetrons for total jamming of enemy's airforce navigation systems," and "long range remote control circuits using power amplifiers," specifying: "For the long range remote control circuits, we mean very long range, more than 200 km..." On March 23, the administrator listed a project which would require an effort from participants in the U.S. Along with the instructions, a YouTube video titled "The Al-25 turbofan engine, part 1" is included.[145]

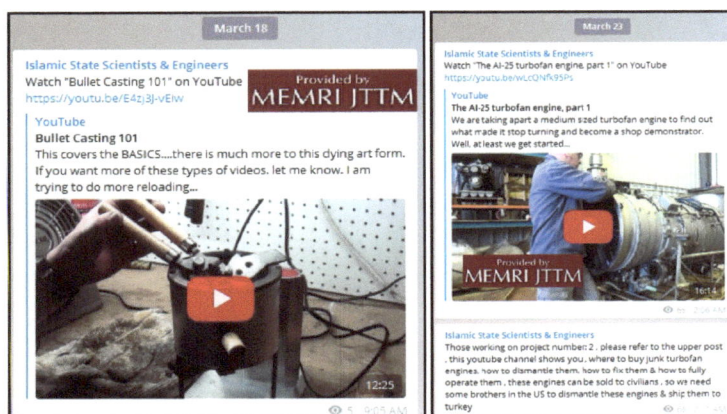

Pro-ISIS Engineers And Scientists Collaborate On Military Projects In Telegram Channel: Part Two, July 21, 2016

The Telegram channel called Islamic State Scientists & Engineers, created in February 2016, continues to operate, seeking to "exploit some channel members' situation to do research for the military benefit of the caliphate" and "collect as much caliphate scientist & engineers as possible from around the world & introduce them to each other." On the channel, individuals with varying professional scientific backgrounds select a project for which their abilities are best suited, and then privately message one another on Telegram to exchange information and work together. Two topics discussed at length in the group have been the manufacture of Tomahawk turbofan engines and GPS-guided missiles. On April 17, the group's administrator wrote: "To join this channel, you must be a muslim engineer, scientist, machinest, or metal worker & you choose to work on one of the projects below & gain knowledge on how to completely achieve them using civilian resources: 1- Manufacturing a chemical reactor for the mass production of nitric acid. 2- manufacturing of a tomahawk turbofan engine using CNC machines & welding. 3- Converting a car engine like a Subaru boxer or a honda to a drone engine using the viking company conversion kit. 4- GPS guided missile using an Arduino flight controller & a gps module. 5- Manufacturing of VHF, UHF, L &S band high power klystrons for jamming purposes, using CNC machines. 6- mass production of smokeless powder for bullets reloading. 7- Manufacturing a long range manually remote controlled surface to air & ground missile using an RF module & an FPV camera...."

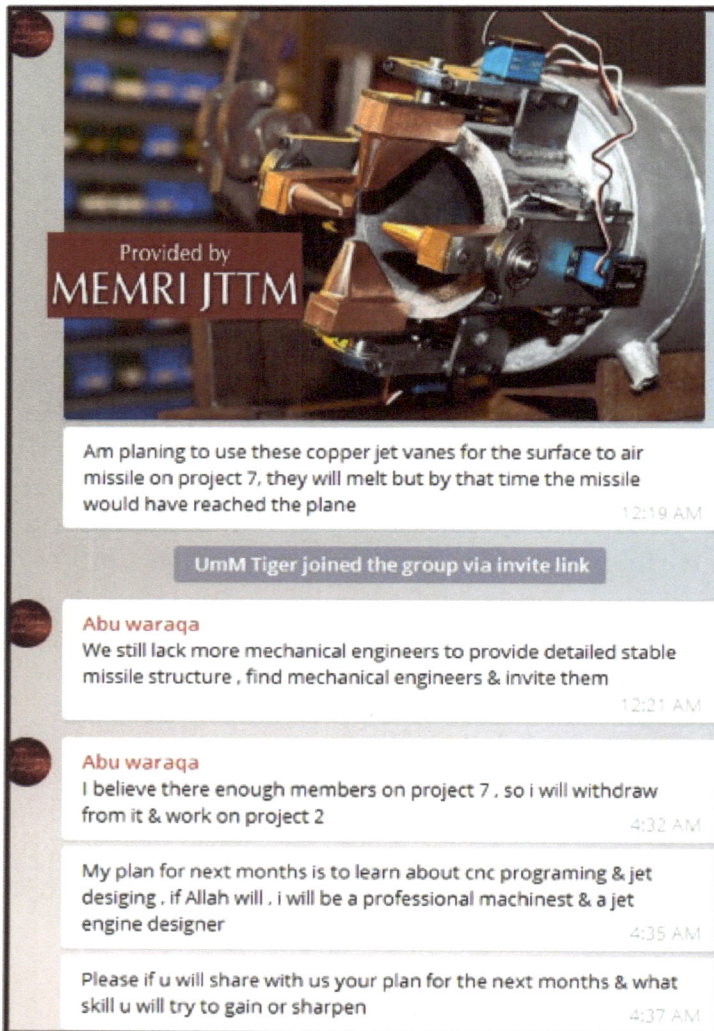

Provided by MEMRI JTTM

Am planing to use these copper jet vanes for the surface to air missile on project 7, they will melt but by that time the missile would have reached the plane — 12:19 AM

UmM Tiger joined the group via invite link

Abu waraqa
We still lack more mechanical engineers to provide detailed stable missile structure , find mechanical engineers & invite them — 12:21 AM

Abu waraqa
I believe there enough members on project 7 , so i will withdraw from it & work on project 2 — 4:32 AM

My plan for next months is to learn about cnc programing & jet desiging . if Allah will . i will be a professional machinest & a jet engine designer — 4:35 AM

Please if u will share with us your plan for the next months & what skill u will try to gain or sharpen — 4:37 AM

Announcement On Telegram: ISIS Radio Launches Experimental Website On The Dark Web, July 5, 2016

The Islamic State's (ISIS) official radio station, Al-Bayan, is currently testing a website to broadcast its content on the Dark Web. For several months now, Al-Bayan's content has been accessible on multiple websites on the Clearnet – the "visible" part of the Internet, which is accessed directly from a user's browser. However, the broadcast on those websites was often interrupted. The announcement of the Dark Web broadcast was posted on Al-Bayan's Telegram channel on July 4, noting that the new webpage is currently experimental. The experimental site's URL ends with onion.to, which utilizes the Tor2web project, allowing Internet users to access TOR (The Onion Router) webpages without installing the dedicated browser. Using Tor2web is more convenient, as it allows users to access Dark Web content directly from their browser, but reduces their anonymity.[146]

Via Telegram, Jihadi Tech Group Offers Tutorial On Using Secure Messaging App Threema, March 9, 2016

On March 8, 2016, the Electronic Horizon Foundation (EHF), a crowdsourcing group that offers tech-related information for jihadis, published a lengthy tutorial on installing and using the secure messaging app Threema. The tutorial, posted on Justpaste.it and circulated on the EHF's Telegram channel, offers some background information about the app, and provides steps to install the app and manage its various functions.[147]

To Avoid Suspension, ISIS Supporters Launch Seemingly Innocuous Telegram "Sports News" Channel – For Promoting ISIS Channels, October 14, 2016

On October 13, 2016, ISIS supporters launched a "Sports News" channel on Telegram. The seemingly innocuous channel, which features a soccer ball image as its avatar and promises to "deliver to you sports news and what's going on in the playing arenas for free," is in fact promoting ISIS channels on the platform. The public channel garnered over 500 members within its first 24 hours of operation. Public Telegram channels have a permanent URL and are searchable via the app. So far, the channel has been acting exclusively as a promotor of other public and private ISIS and pro-ISIS Telegram channels, but has not posted any original content. ISIS supporters have used this strategy of giving accounts innocent-seeming names on other platforms, such as Twitter and Facebook. However, this strategy is not always effective in preventing suspension, since once the reader gets past the profile description and innocent images, the content generally reveals the account's true purpose. Other ISIS supporters have taken a subtler and more covert approach – not only creating innocuous profiles but also refraining from clearly showing support and sympathy for the Islamic State.[148]

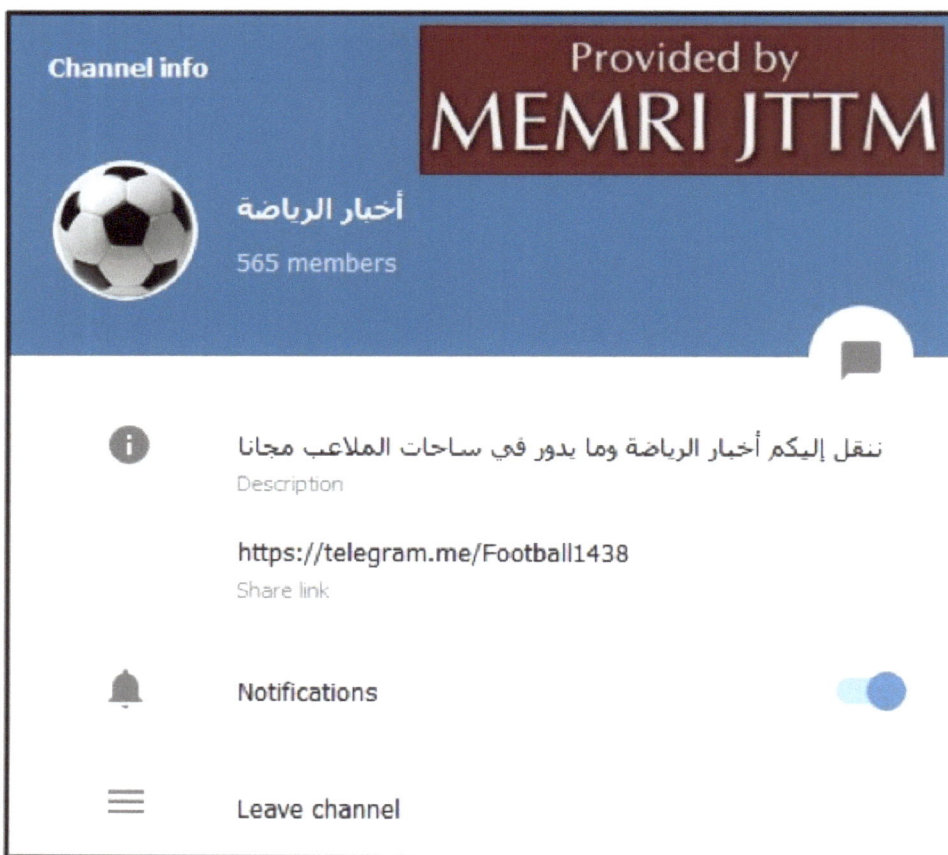

"Sports News" Telegram profile

Sports News posting link to A'maq's Telegram channel and reposting a link to a channel belonging to Nasher News, a major disseminator of ISIS content

ISIS Supporters Tighten Security Measures To Join ISIS Channels On Telegram, January 6, 2016

ISIS supporters are doing more to stop their channels on the secure communication app Telegram from being reported and shut down. One pro-ISIS blog is now only allowing people to join the English Nashir channel, a top disseminator of ISIS content in English on Telegram, by submitting their request online. Requests, it says, are generally processed within 24 hours.[149]

ISIS Supporters Move Conversation Offline, Create Chatrooms On Telegram, August 25, 2015

On August 25, 2015, two ISIS supporters on Twitter announced the creation of an ISIS chatroom on Telegram. In light of arrests, and leaked information, both fighters and prospective members and supporters have recently taken more precautions with their discussions on social media and apps.[150]

Pro-ISIS Telegram Channel Warns ISIS Supporters Against Using WhatsApp, August 11, 2016

On August 11, 2016, the pro-ISIS Telegram channel "Nasher Graphics" posted a message to ISIS supporters, warning them to refrain from joining groups on the messaging service WhatsApp that claim to be pro-ISIS. The post reads: "We warn you against using WhatsApp, especially the brothers in European countries and the Gulf states, because it is an app that is heavily monitored and exposes the brothers to danger." Referring to an announcement on an alleged pro-ISIS group on WhatsApp that featured a link to another group that it said is falsely using the name of the known pro-ISIS Ashhad Media Foundation, likely as a way to fool ISIS supporters (see right side of image below), the message stated: "We think that the group that shared this [link] is a channel that took the name of [the pro-ISIS media group] Ashhad. Therefore – beware, beware, brothers. We ask Allah to protect the monotheists everywhere from any harm." It added: "We warn you, oh supporters, against using WhatsApp in your [media] support [activities], since it may expose you to arrest. We warn you against these types of evil channels that publish phone numbers. Our experts looked into the matter and it has become evident that the program is heavily monitored."[151]

Pro-ISIS Group Drafts Tweets For Followers To Post Under Hashtags On Twitter, October 24, 2016

On October 24, 2016, the pro-ISIS Telegram group Ash-had Brigade posted a message stating that it "focuses on drafting tweets and attack hashtags on #Twitter." It added, "We draft [the tweets] and you post them, oh media men of the caliphate."

Pro-ISIS Hacking Groups – Releasing "Kill Lists" Of U.S. And Western Officials, Military, Law Enforcement, And More

Pro-ISIS Hacking Group Caliphate Cyber Army (CCA) Promotes Latest Video Release On Telegram, Joins Forces with Pro-Palestinian Hackers AnonGhost, January 11, 2016

On January 8, 2015 the Caliphate Cyber Army (CCA), a group of pro-ISIS hackers, circulated a link to their latest release, "The Establishment of Caliphate Ghosts," that it had uploaded to the Internet Archive. The 13-minute video showed off some of the CCA's successful hacks, and featured the hacking of the website of the Global Security & Training Services (GSTS), an Israeli company. The video features the usual ISIS rhetoric of threats to the West, the conquering of cities, and the destruction of landmarks such as the White House, Big Ben, and the Eiffel Tower. This group of hackers also promoted the name "Caliphate Ghost" in the video; this refers to Pro-Palestinian hackers AnonGhost with whom the CCA have joined forces. The video features stills showing the GSTS website being hacked, and displays a few articles reporting on various websites they have hacked.[152]

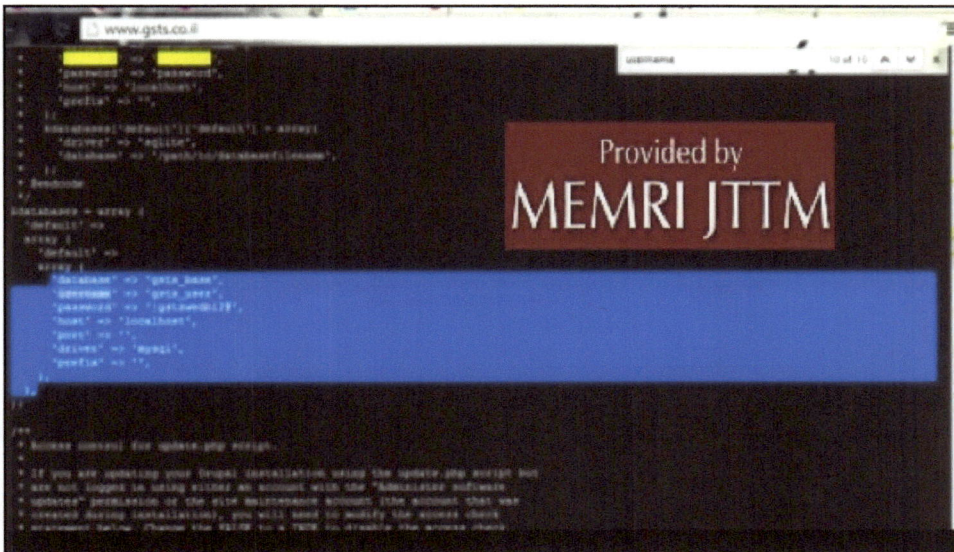

CCA Launches New Telegram Channels, February 4, 2016

On February 3, 2016, the Cyber Caliphate Army (CCA) created two new channels on the secure messaging app Telegram. The first channel is for the Cyber Caliphate Army itself. The second channel is called "Sons Caliphate Army."

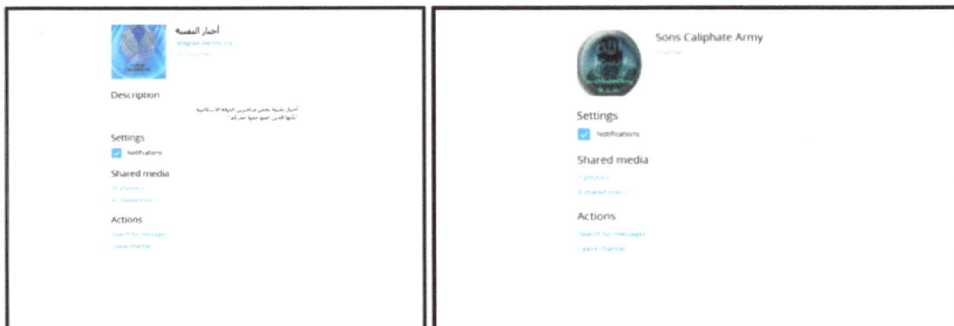

A post on this second channel claimed that the Cyber Caliphate Army had hacked 15,000 accounts, one of them the Independence Hall Tea Party PAC in the Delaware Valley area. [153]

CCA Video Made With "Cute Cut" For Apple iPad, Shows Group Editing Videos, Reiterating Oath Of Loyalty To ISIS, March 14, 2016

On March 14, 2016, the Cyber Caliphate Army (CCA) posted a video to its Telegram channel showing members of the group editing video using the "Cute Cut" editing software for Apple iPad. The video featured a reiteration of the group's oath of fealty to ISIS leader Abu Bakr Al-Baghdadi.[154]

CCA Leaks New Jersey And New York Police Officer Data On Telegram, March 7, 2016

On March 2, 2016, the pro-ISIS hacker group Caliphate Cyber Army posted a statement on its Telegram channel claiming to have obtained personnel data of 55 police officers from New Jersey and New York. The announcement included a link to the data, which contains the names, ranks, phone numbers, work addresses and working locations of the police officers. The link leads to an Arabic-language file-sharing website where users can download data in the form of an Excel spreadsheet.5[157]

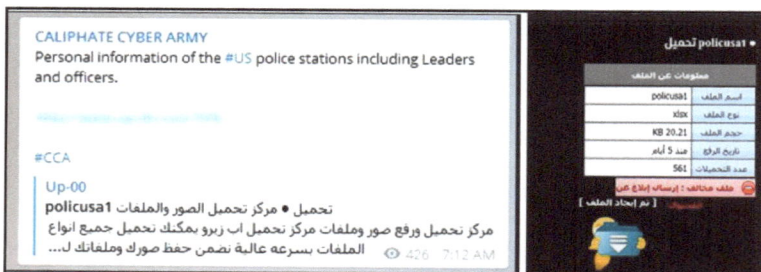

CCA Calls To Celebrate Brussels Attacks Using Trending Hashtags, March 22, 2016

Following the March 22, 2016 Brussels attacks, the Caliphate Cyber Army (CCA) posted messages on its official Telegram channel calling on ISIS supporters to "storm" trending hashtags on social media with celebratory posts regarding the attacks. The group wrote: "NOTICE: Please take part in this online battle in twitter and storm the trending hashtags we provided you with any information from dawlah [ISIS]. Be it a video, images, news etc. Post and post and you have your ajr [reward] from Allah." The hashtags provided were #Brussel, #Brussels, #Bruxelles, and #Brusselsattacks.[156]

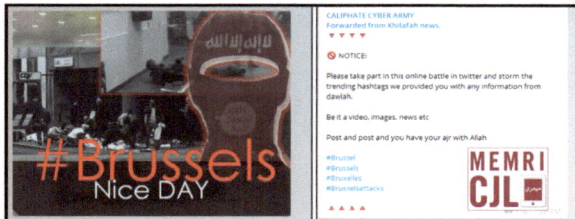

CCA Leaks Details On U.S. Police On Telegram, March 14, 2016

On March 14, 2016, the Caliphate Cyber Army posted info on U.S. police officers on its Telegram channel. This is the third time in the past two weeks that CCA has published names and details of American police officers.[157]

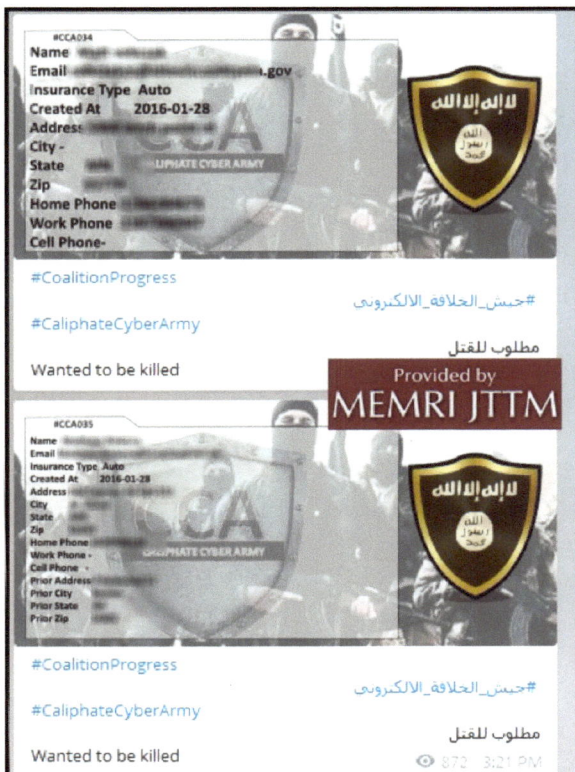

After Publishing State Dept Kill List, CCA Telegram Channel Is Shut Down; New Channel Launched Promising More Hacks, Mocking FBI, May 3, 2016

After leaking the personal info of U.S. State Department employees on April 24, 2016,[158] the pro-ISIS hacking group Cyber Caliphate Army's Telegram channel was shut down. It is unclear whether the group closed its own channel or whether it was shut down by Telegram administrators. However, on May 3, 2016, Caliphate Cyber Army launched a new telegram channel, and posted messages mocking the FBI and promising further hacks.[159]

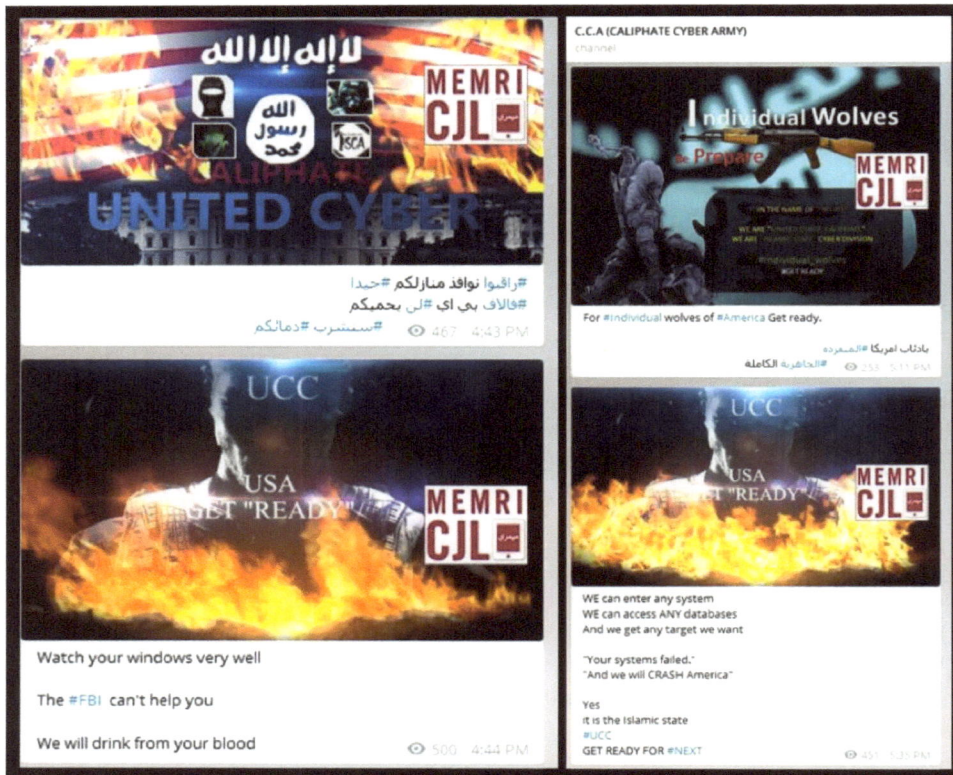

CCA Back On Telegram, Announces Formation Of New Collective "United Cyber Caliphate," April 4, 2016

On April 4, 2016, the Cyber Caliphate Army (CCA), whose Telegram account had been shut down, returned to the platform. The CCA announced a cooperative with like-minded cyber jihadi groups in order to "expand... operations," as well as the launch of additional account titled United Cyber Caliphate (UCC) which was to be a collective Telegram account for four cyber jihadi organizations – Ghost Caliphate Section, Sons Caliphate Army, Caliphate Cyber Army, and Kalachnikv E-security team.[160]

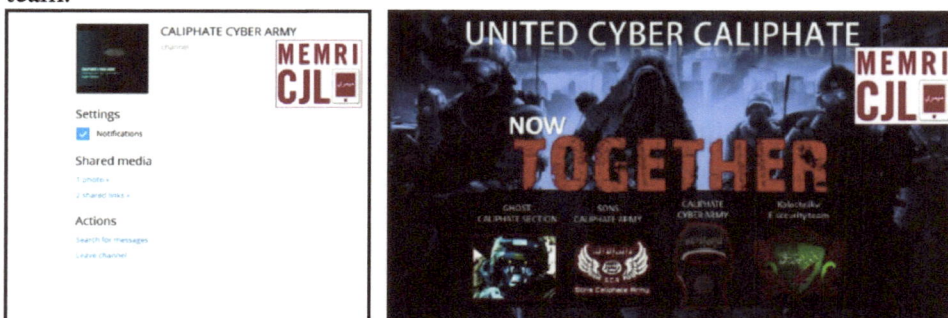

CCA Threatens U.S. With Upcoming Attack, April 21, 2016

On April 20, 2016, the pro-ISIS hacking group Caliphate Cyber Army posted a message on its Telegram channel stating that the U.S. would be its next target. The caption accompanying the message read: "Remember the challenge soon. Wolves individual processed in the heart of America strike will be in sha Allah." The same threat was circulated on a sister channel, United Cyber Caliphate, in both English and Arabic; a photo of what appears to be ISIS decals on a car window, with two U.S. flags in the background indicate that this supporter or operative is currently on U.S. soil. The threat was also circulated on the Telegram channel belonging to Kalachnikv E-security Team.[161]

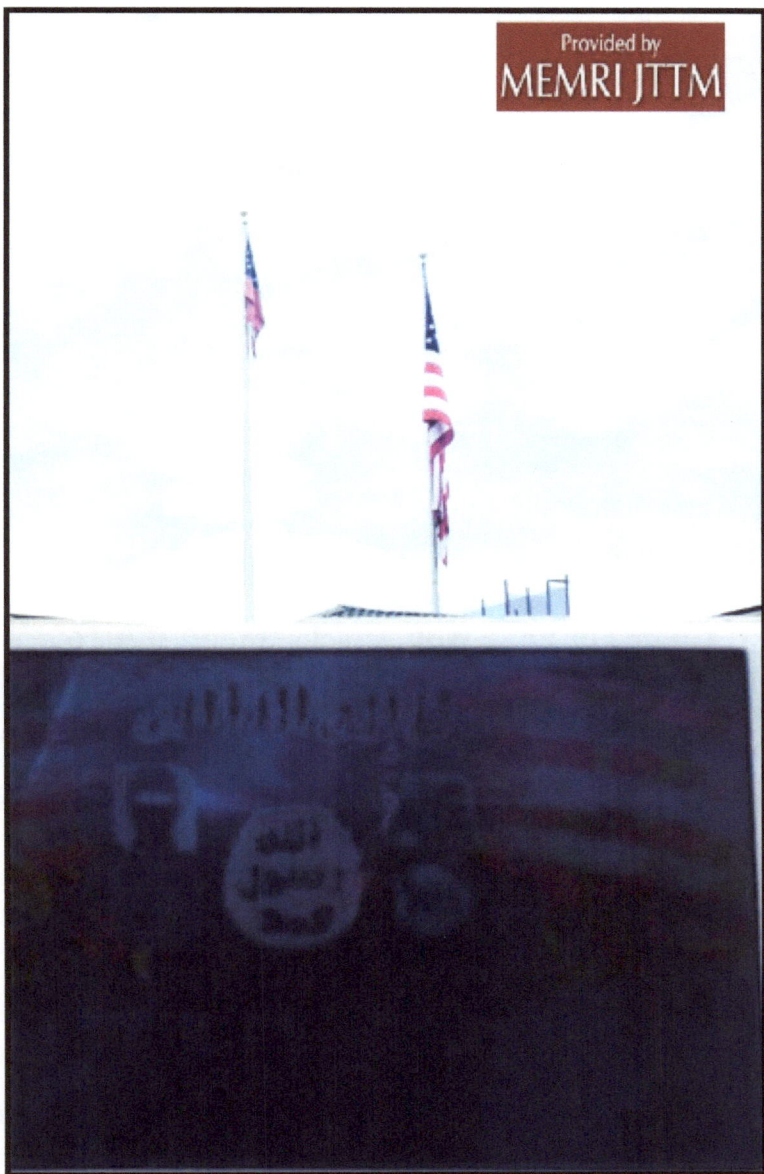

CCA Circulates Warnings About Fake ISIS Apps "Aimed At Infiltration," June 1, 2016

On June 1, 2016, the Caliphate Cyber Army (CCA) posted a notice stating that fake versions of the ISIS Al-Bayan Radio, A'maq news agency, and other apps were being "distributed online, with the distributor claiming availability in several languages." The notice went on to warn, "This is clearly aimed at infiltration. We advise all supporters of the Khilafa to rely on official channels to download these apps, as well as to verify the checksums before use." The same day, "Amriki Muhajer" also posted on Telegram, "Warning: Dubious sources published a fake version of the Amaq Agency Android app, aimed at breaching security and spying. We advise to avoid downloading any app, except via the official Amaq channels and recommend to verify with the officially published checksums before installation."[162]

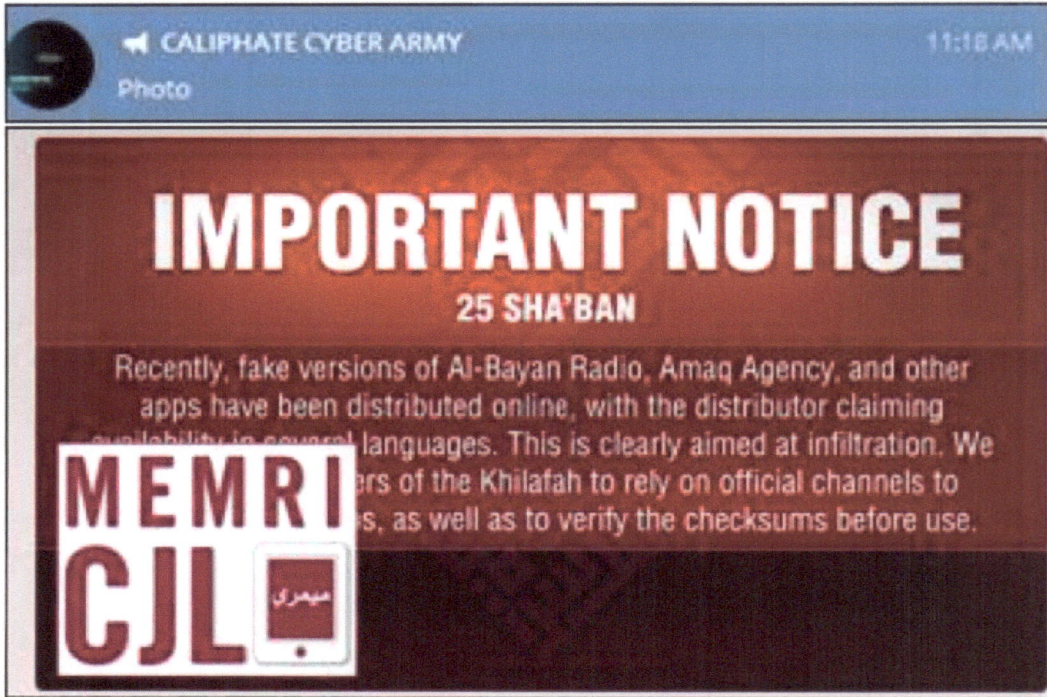

"On Telegram, Pro-ISIS Hacking Group CCA Posts Guide For Lone Wolf Attacks, June 23, 2016

On June 22, 2016, the pro-ISIS hacking group Caliphate Cyber Army posted a graphic with ways of carrying out lone-wolf attacks. Some of the locations it suggested for attacks were public places such as markets and restaurants, and other instructions concerned attacking targets inside their homes. The graphic also provided various ways to carry out the attacks, including poisoning, stabbing, shooting, and throwing rocks. It even suggests screaming at someone as a form of an attack.[163]

CCA Posts "Lone Wolf Scorecard" On Telegram, Marks Recent Stabbing, Bombing, Vehicular Attacks; Shooting, Poisoning, Beating Attacks Not Yet Checked Off, July 26, 2016

The Caliphate Cyber Army (CCA) posted an image on its Telegram channel featuring various methods for lone wolf attacks. The graphic includes shooting, stabbing, bombing, vehicular attacks, rock throwing, poisoning, and beating. After the July 14, 2016 vehicular attack in Nice, France, CCA reposted the scorecard, but this version had a red circle around the vehicular attack image. The post included the text: "Mission has been completed. Waiting next." Following the July 19 stabbing attack in Wurzburg, Germany, CCA once again updated its scorecard to include a stabbing attack. The post included the text "Second operation successfully completed. Waiting for the next. Islamic State win." CCA updated its scorecard once again on July 25, following the suicide bombing in Ansbach, Germany. This time the bombing icon was also included, along with the text "Ansbach, Germany. Operation completed. Alhamdullilah." On July 26, the scorecard was updated yet again to include two stabbing attacks following the killing of the priest that day in Normandy, France. The new post included the text: "Normandy. Mission has been completed successfully. Waiting for the next. UCC. CCA."[164]

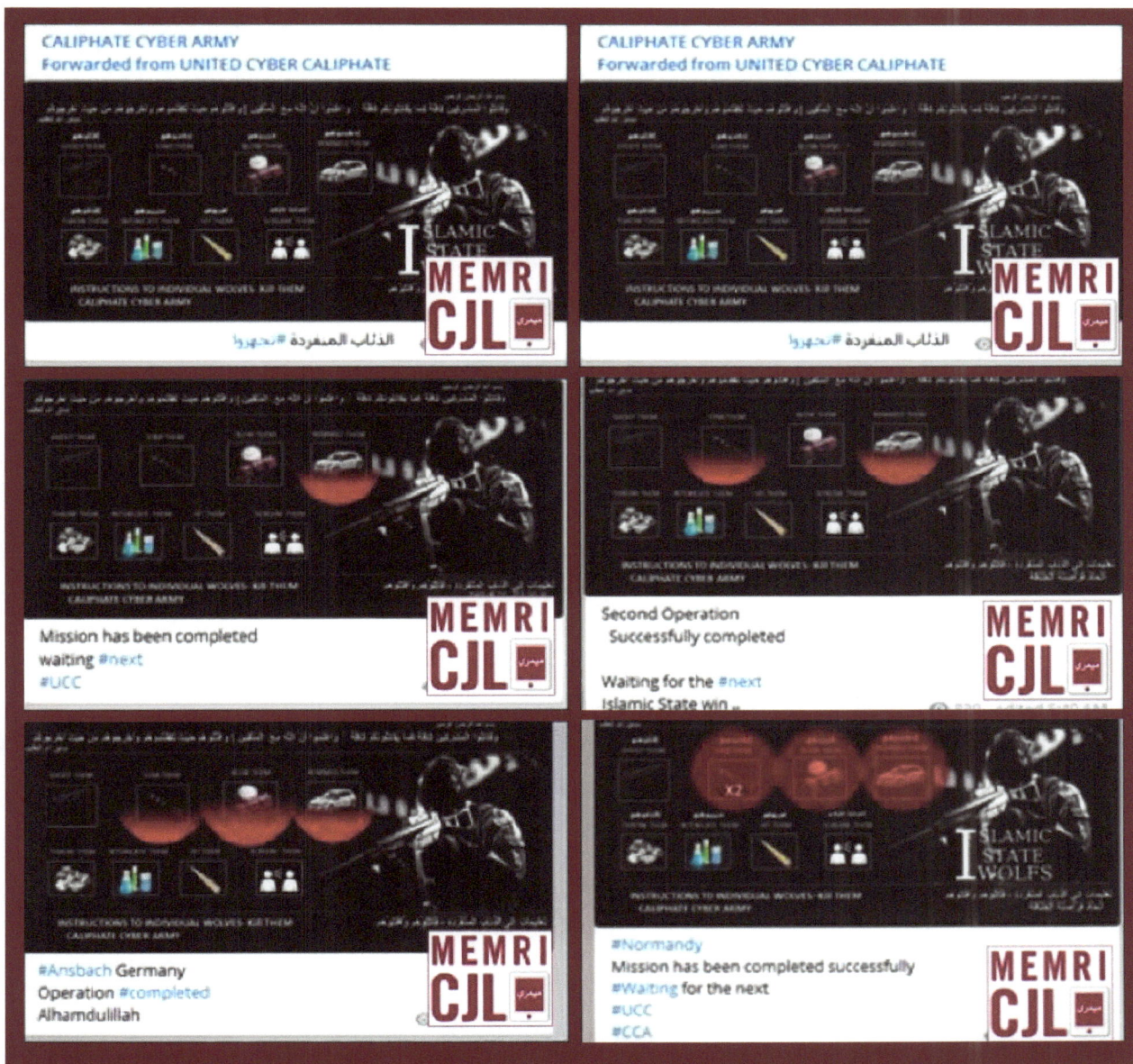

CCA Releases List Of 3,247 Individuals Arrested In Dallas 2014-2016; Leaks List Of What It Claims Are Police Salaries, July 8, 2016

On July 8, 2016, the Caliphate Cyber Army (CCA) released on Telegram a list of 3,247 individuals, both men and women, arrested in Dallas, Texas during 2014-2016, most of them members of minority groups. The list includes other details such as location of arrest, and whether the individual arrested was armed.[165]

Dallas arrest list released by CCA

CCA, UCC, Ghosts Caliphate Groups Post "Kill List" Featuring U.S., Australian, Canadian Citizens, June 7, 2016

On June 7, 2016, the pro-ISIS hacking groups Caliphate Cyber Army and its affiliates United Cyber Caliphate and Ghosts Caliphate posted a list of hundreds of names of citizens of the U.S., Australia, and Canada. Accompanying the list was a graphic image that read "Caliphate 'Ghosts' Caliphate Cyber Army/ All World Cant Stop Islamic State/ You are gathered against us./ 'We will kill you all (together).'" The comment on the post read, "Fight the Cross.' Following them. Kill them. Strongly' Revenge for Muslims."[166]

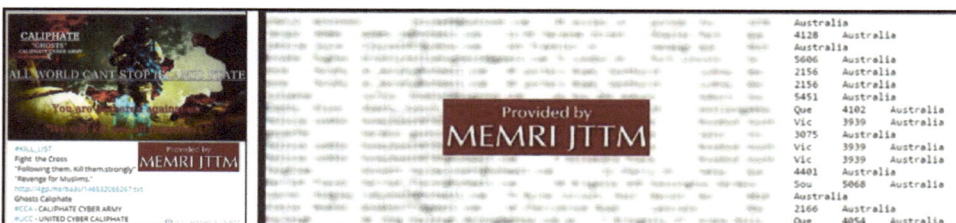

CCA Releases Kill List, Personal Contact Information Of USAF Air Mobility Command – AL, HI, NJ, PA, TX Officials Targeted, June 21, 2016

On June 15, 2016, the details of 11 personnel of the U.S. Air Force Air Mobility Command were leaked by the pro-ISIS hacking group Caliphate Cyber Army (CCA). The list contains their phone numbers, email addresses, command ranking, base names, cities, states, and street addresses. The Telegram post reads: "Wanted to kill 'immediately' #USA Air Mobility Command Personal Data Hacked #CCA #UCC Islamic State."[167]

CCA Releases Kill List With 1,693 Names Of "Crusaders And Jews" In U.S., July 5, 2016

On July 3, 2016, Caliphate Cyber Army (CCA) released via Telegram a kill list containing the names of 1,693 "crusaders and Jews" in U.S. cities including Chicago, San Diego, Phoenix, San Antonio, Boston, and Colorado Springs. In addition to personal details such as emails and home addresses, the kill list also included the churches or synagogues purportedly attended by those listed. After the content in the original link was removed, the CCA re-posted a batch of new working links on July 5, 2016 on its Telegram channel.[168]

CCA Claims To Have Hacked Official Saudi Government Portal And Obtained Official Government Database, July 5, 2016

On July 4, 2016, the Caliphate Cyber Army (CCA) announced on Telegram that it had successfully hacked the main Saudi government portal and obtained a database with the names, email addresses, and job listings of all Saudi government officials. The Caliphate Cyber Army announcement included a partial list of several dozen Saudi government officials in local municipalities, public sector companies, mail workers, university staff, and more. The hacking group claimed that this was just a small sample of the database it had obtained.[169]

CCA Links To New "OPSEC IT" Telegram Channel, September 22, 2016

On September 18, 2016, the Caliphate Cyber Army (CCA) circulated a link to a new Telegram channel for operational security in the realm of information technology. Named "OPSEC IT," the channel is intended "to train [supporters] to use the basic security precautions that need to be taken while online," including "what apps to use and how to use and tutorials on Tor, PGP, Tails OS, Bit Message [sic] and a few other advanced softwares [sic] as well." As of September 20, 2016, content posted in the OPSEC IT channel included a recently distributed guide on how to use the encrypted messaging app ChatSecure.[170]

CCA Releases List, Satellite Images Of U.S. Military Bases, June 7, 2016

On June 7, 2016, the Caliphate Cyber Army (CCA) and its affiliates United Cyber Caliphate (UCC) and Ghosts Caliphate posted a list of U.S. Air Force bases, along with several satellite images, including of F.E. Warren and MacDill bases, on their Telegram channels. Accompanying the information was an image with the text: "Caliphate 'Ghosts' Caliphate Cyber Army. All World Cant Stop Islamic State. You are gathered against us. 'We will kill you all (together)." The message accompanying the post read: "Kill the cross. 'Revenge For Muslims.' Wholly US military airfields burnt it."[171]

CCA Promotes Upcoming Operation On Telegram, January 5, 2016

On January 5, 2016, the Caliphate Cyber Army posted a message on its Telegram channel promoting an upcoming operation, which they promised would be a "heavy caliber surprise."[172]

United Cyber Caliphate (UCC) Targets Australian Websites, April 14, 2016

On April 14, 2016, the United Cyber Caliphate, formerly known as the Caliphate Cyber Army, announced on Telegram that Australia was the next target for their cyber-attacks. Following this announcement the group posted a list of Australian web-sites that the group hacked.[173] The formation of the UCC was announced in April 2016 by the CCA, which described it as a cooperative of cyber jihadi groups that aimed to "expand... operations," including Ghost Caliphate Section, Sons Caliphate Army, Caliphate Cyber Army, and Kalachnikv E-security team; the UCC Telegram account was launched at the same time.[174]

In Response To Twitter's Crackdown On Extremist Accounts, UCC Alleges It Hacked 5,000 Accounts On The Platform, August 23, 2016

On August 22, 2016, the United Cyber Caliphate (UCC), a pro-ISIS hacking group, claimed that it had hacked 5,000 new Twitter accounts in addition to the 5,000 that it previously claimed to have hacked. UCC claims that it is hijacking these hacked accounts in order to further spread its propaganda. Via its Telegram channel, the group hinted that this new wave of hacks was a response to Twitter's August 18 announcement that it had suspended over 200,000 extremist accounts on its platform over the past six months.[175]

UCC Releases Kill List With Details Of Over 2,000 Officers And Soldiers On U.S. Military Bases, July 20, 2016

On July 20, 2016, United Cyber Caliphate (UCC) posted a list with what it said was the names and personal details of over 2,000 officers and soldiers serving in U.S. military bases on its Telegram channel. The image attached to the post read: "Officers and soldiers serving in the US military bases. We want them #dead. #Revenge for Muslims." Hours before the list was posted, the Caliphate Cyber Army (CCA), a UCC affiliate, warned that revenge would be sought for the 160 civilians killed by "Crusader" aircraft in Manbej, Syria. The entry reads: "160 civilians were killed in 'Manbej' because of the Crusaders aircraft and we did not see any 'dog' spoke about them. And all media shut up... 'You are cowards.' Swear to 'allah.' We will Kill you all. Your wives. Your children. Even a baby. We will slaughter you. For Islamic State. #Revenge #Soon_Crusaders."[176]

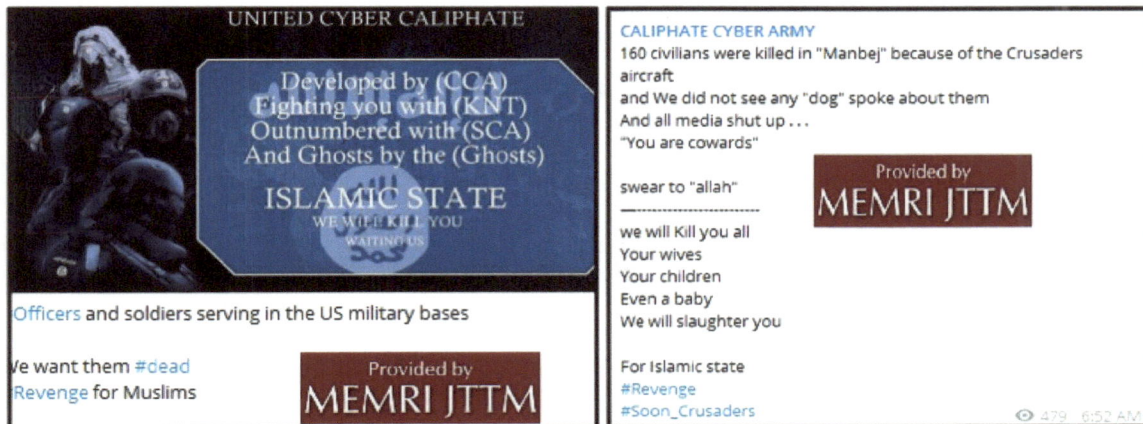

UCC Distributes Request For Scientific Personnel On Telegram, August 31, 2016

On August 29, 2016, the United Cyber Caliphate (UCC) Telegram channel forwarded a message from the Caliphate Cyber Army (CCA) Telegram channel seeking individuals with scientific capabilities to assist in their mission.[177]

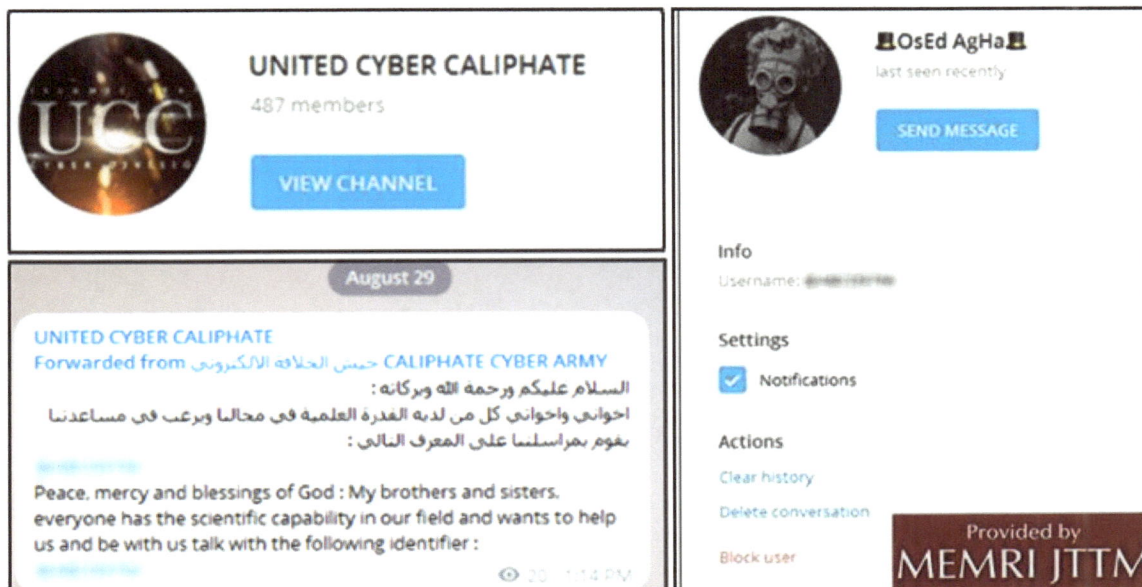

UCC Leaks List Of Allegedly Prominent NYC Citizens, April 21, 2016

On April 21, 2016, the United Cyber Caliphate Cyber released, on its Telegram channel, a list of the names of over 3,000 of "the most important citizens of New York and Brooklyn." The graphic threatened: "We want them dead, shut them down." The list appears to include primarily the contact details of New Yorkers; however, some people on it are from Kansas, Michigan, and other places.[178]

UCC Releases Personal Details Of U.S. State Department Employees, April 25, 2016

On April 24, 2016, the United Cyber Caliphate announced on its Telegram channel that it had hacked the U.S. State Department and released a list of personal details of some of its employees. The statement is titled "wanted to be killed" and reads: "USA You are our primary goal. Your system failed to Tackling [sic] our attacks. Now we will Crush you again."[179]

Sample of leaked details (personal information redacted by MEMRI)

UCC Says It Hacked Database Of Saudi Defense And Interior Ministries, Releases Info Of Security Officials, April 25, 2016

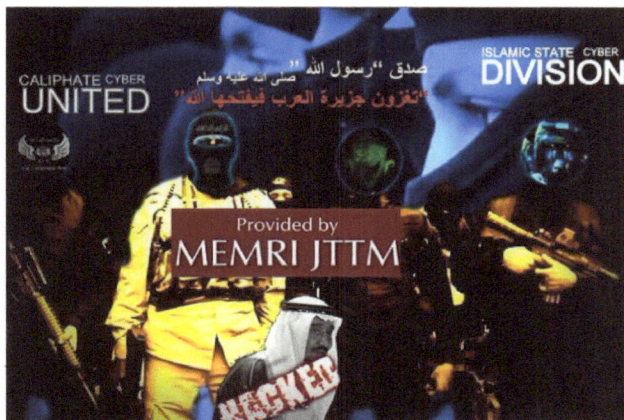

On April 22, 2016, United Cyber Caliphate (UCC) released, on its Telegram channel, names and personal details it said belonged to of Saudi security officials that it claimed to have obtained by hacking into the database of the Saudi Defense and Interior Ministries. UCC stated that it had gained access to the details of 18,000 employees in the security departments, but released only a few screenshots of personal information pages showing names, ages, nationalities, ID numbers, and so on. Most of the screenshots were captioned with threats to the Saudi government and a message urging lone wolves to carry out attacks. One such message read: "Do not dare think about ISIS as we would only respond to you by death." Another urged ISIS supporters to phone the individuals whose details were released so that they could run before they face death.[180]

UCC Hacks Vancouver Florist, May 11, 2016

On May 9, 2016, the United Cyber Caliphate announced on its Telegram channel that it had hacked the website of a Vancouver florist, Floral Revelry, and leaked its database online. The group's choice of target could indicate that it is not engaged in serious cyber-jihad activity.[181]

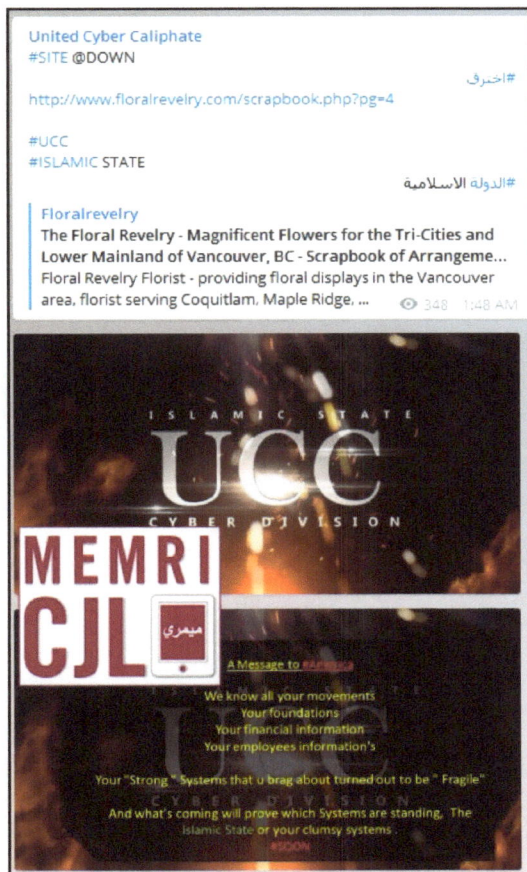

UCC Posts On Telegram Announcement Of Hack Of "Arkansas Library Database," May 27, 2016

On May 26, 2016, the United Cyber Caliphate posted, on its Telegram group, an announcement by the Caliphate Cyber Army that it had hacked the "Arkansas Library DATABASE." Another post, in Arabic, noted: "Caliphate Cyber Army/Society of the Libraries in Arkansas/Entirely hacked/The Site Database."[182]

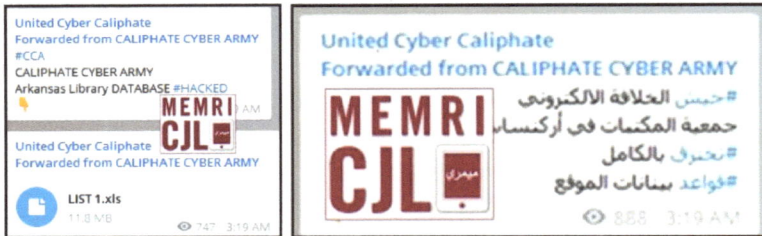

UCC Posts Social Media Safety Instructions, May 25, 2016

On May 25, 2016, the pro-ISIS hacking collective United Cyber Caliphate posted a series of safety instructions for "brothers and sisters" on social media on its Telegram channel. The instructions include:

1. Only accept content from trusted accounts.

2. Never share personal information such as pictures or voice recording, and if you must do so, delete it immediately afterwards.

3. Hide all account information such as email address, phone number, etc.

4. Use Tor or a VPN when connecting to social media.

5. Never give out your account to any strangers, even ISIS supporters.

6. Never open links from untrusted sources.

The post concludes by saying: "Finally, as you know, everyone is trying to find any information concerning you or your location. Don't underestimate these advices, it is for your own protection."[183]

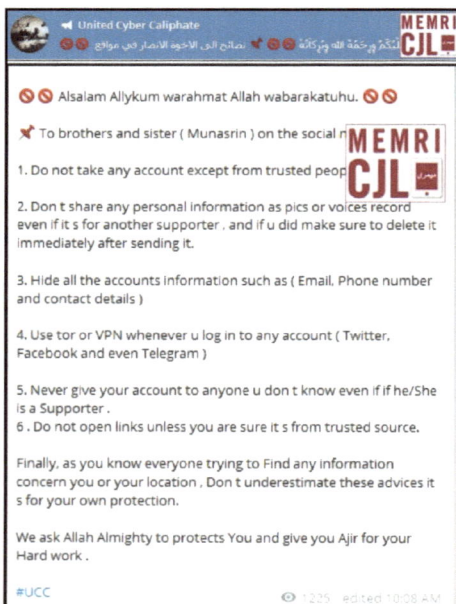

UCC Releases "Kill List" With 4,681 Names From Around The World – Including U.S., China, India, Australia, U.K, Canada – And From Microsoft, IBM, Walmart, Home Depot, Oracle, Yahoo, ExxonMobil, June 22, 2016

On June 21, 2016, United Cyber Caliphate (UCC) released, on its Telegram channel, a "kill list" with the personal information of 4,681 individuals from the U.S., China, Germany, India, Australia, the U.K., Canada, and South Korea. The list included employees from large corporations, including Barclay's Bank, Walmart, Amazon, IBM, Haliburton Energy Services, Microsoft, Samsung, ExxonMobil, Home Depot, Oracle, Sprint Corporation, Shell International E&P, U.S. General Services Administration, JP Morgan Chase, Rosettanet, Michigan State University, Lockheed Martin, Georgia State University, Nordstrom, FedEx Corporation, and Yahoo. The UCC graphic for the list reads: "O individual wolves out there in the world Kill the Cross wherever you find it. For there 'war on Islam' and 'mujahideen.' Kill them strongly... Kill them hardly #Kill them Revenge for Muslims #World Get ready for the second round." The kill list was also forwarded on Telegram by the ISIS-affiliated hacking group Kalachnikov E-Security Team.[184]

UCC Kill List Targeting Canadians, Containing Over 12,000 Entries, June 28, 2016

On June 28, 2016, the United Cyber Caliphate released, on its Telegram channel, a kill list of Canadians, with over 12,000 entries, though some are duplicates. The Telegram post included an image with the text: "Individual wolves in Canada: A list of 12,000 crusader. Kill them immediately. Enter them home and slay." The image is accompanied by the text: "#Targeting direct #Canada It wanted to be killed immediately 12,000 Crusader Slay them."[185]

UCC Releases Kill List Of 289 U.S. Army Corps Of Engineers Personnel, July 20, 2016

On July 19, 2016, United Cyber Caliphate (UCC) released a kill list on its Telegram channel containing what it said was personal details of 289 members of the U.S. Army Corps of Engineers. Most of the personnel listed are from Oklahoma; however, there are dozens of names of a handful of other states. The message accompanying the information stated: "Wanted to kill. KILL THEM ALL. U.S. Army Corps of Engineers. #HACKED."[186]

UCC Releases Kill List Of Over 700 U.S. Army Personnel, July 25, 2016

On July 25, 2016, the United Cyber Caliphate released a kill list of over 700 U.S. Army personnel on its Telegram channel. The list included the message, "We want them dead. Revenge for Muslims. Kill the dogs."[187]

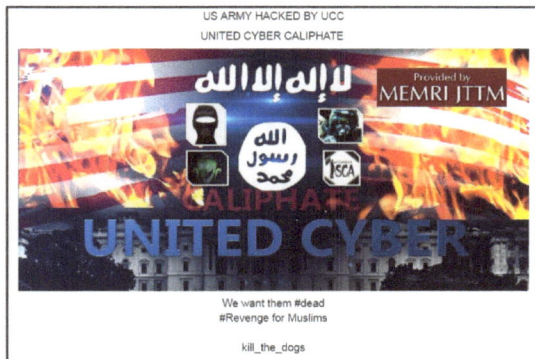

UCC Posts Infographic Highlighting Hacking Prowess, July 29, 2016

On July 28, 2016, United Cyber Caliphate (UCC) posted an infographic on its Telegram channel highlighting the group's most prominent hacks, listing them by countries targeted, the number of websites hacked, and the number of soldiers whose information was leaked.[188]

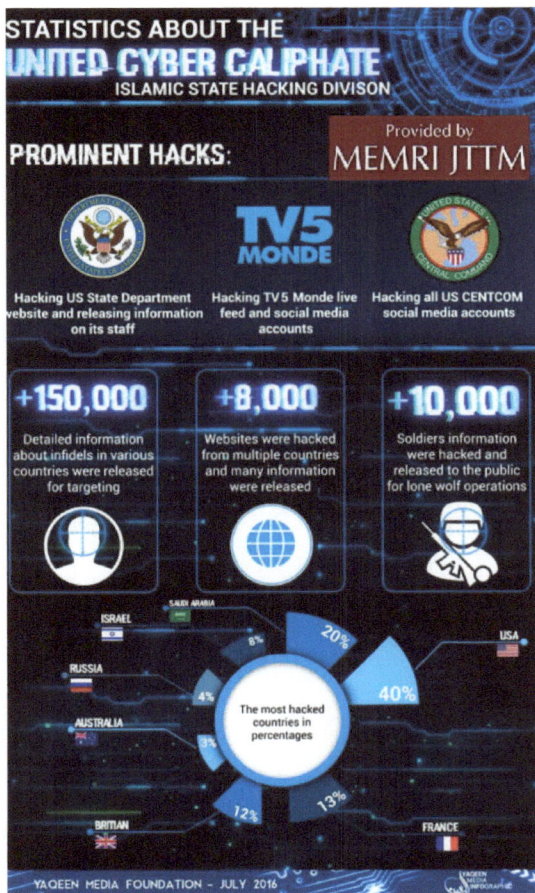

UCC Releases Kill List Featuring U.S. Air Force Personnel, August 3, 2016

On August 2, 2016, the United Cyber Caliphate (UCC) circulated a kill list, via its Telegram channel, featuring the personal details of U.S. Air Force personnel. Those on the list are stationed in bases both in the U.S. and overseas.[189]

UCC Praises Member's Efforts In Wake Of Arrest, Warns Of Perils Of Cyber Jihad, October 24, 2016

On October 23, 2016, USS released a PDF document on its Telegram channel focusing on the recent arrest of one of the group's members. The group uploaded three versions of the document, in Arabic, English and French. The document reveals that the hacker who was arrested is a 26-year-old Kuwaiti citizen. The letter also briefly mentions that three other hackers have been arrested, but their real identities and their fates are still unknown.

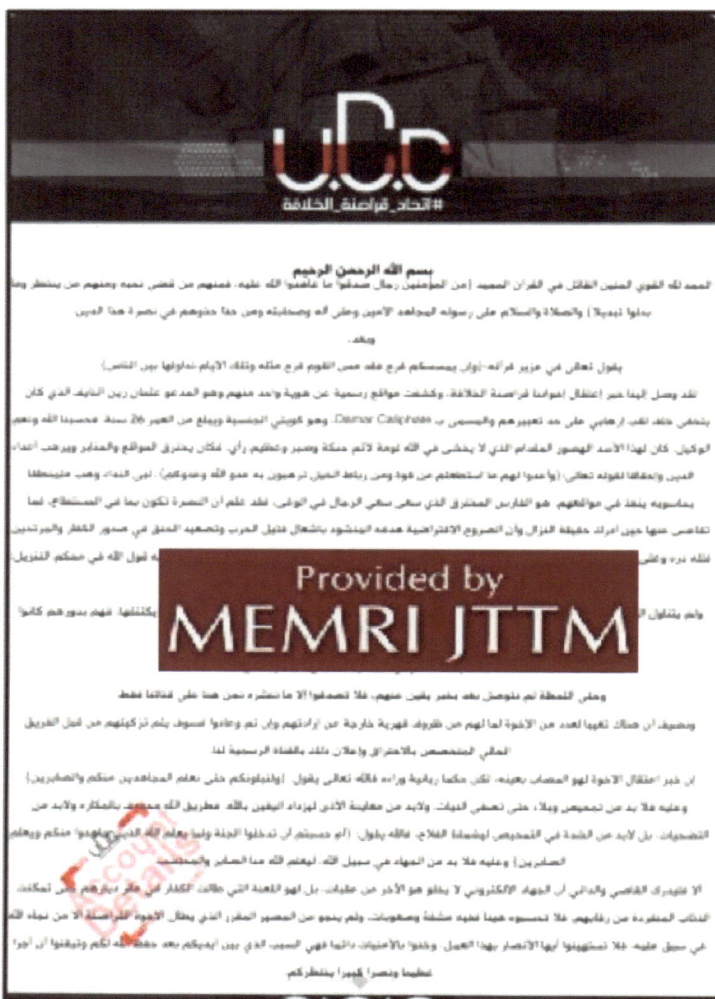

Announcement posted on Telegram

Sons Caliphate Army Shares Information About Hacks, Threats, Jihadi Content On Telegram, March 31, 2016

On February 12, 2016, the Cyber Caliphate Telegram account announced the launch of a new breakaway group called Sons Caliphate Army and provided a link to the new channel.[190]

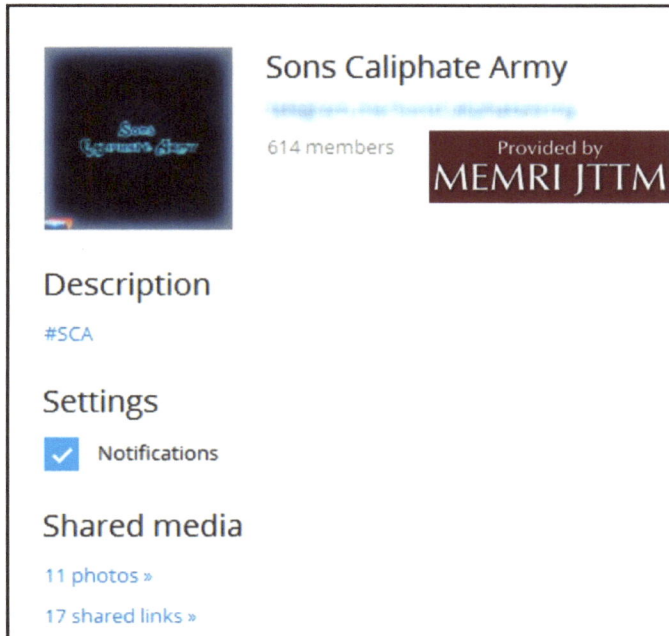

Sons Caliphate Army Telegram Boasts Of Facebook, Twitter Hacks, Vows To Continue Cyber War, June 23, 2016

On June 22, 2016, the Sons Caliphate Army released a video showing Facebook and Twitter accounts that the group claims it had hacked. This video's content echoed that of a previous release published by the group on February 23, 2016, which lauded those partaking in cyber efforts in the name of the Islamic State. Mark Zuckerberg of Facebook and Jack Dorsey of Twitter were mentioned again in this latest release. The new video mocked the two social media platforms' efforts to combat the presence of pro-ISIS accounts and content on them. It also called their removal of pro-ISIS content is hypocritical, since both of them claim to promote democracy and freedom of expression. The video, titled "The Word of the Sword," is approximately 19 minutes in duration.[191]

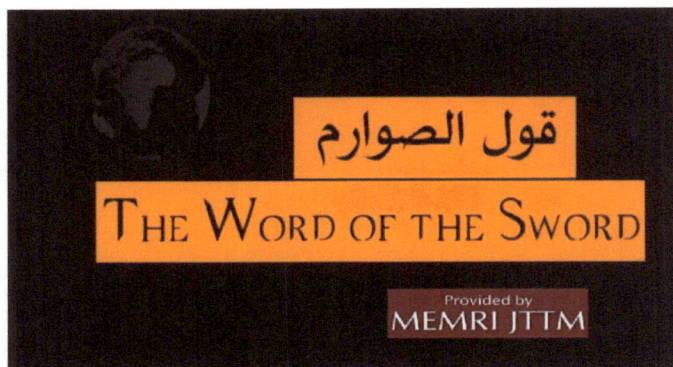

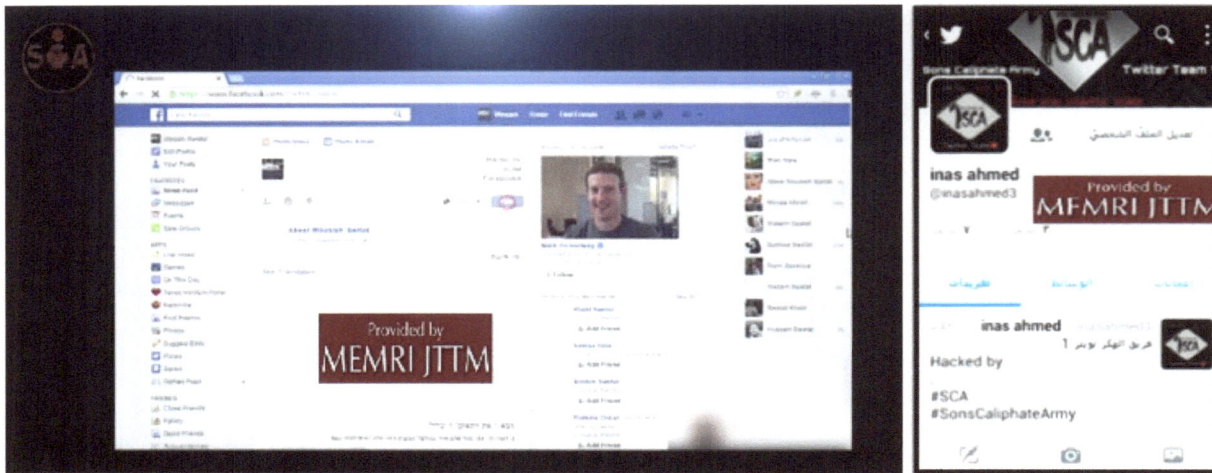

Hacked Facebook account; hacked Twitter accounOn Telegram, Sons Caliphate Army Announces Twitter Accounts It Hacked Have Been Disabled, Mocks Facebook And Twitter CEOs Mark Zuckerberg And Jack Dorsey, September 12, 2016

On September 11, 2016, the Telegram channel of the pro-ISIS hacking group Sons Caliphate Army posted an announcement stating that most of the Facebook and Twitter accounts hacked by the group had been disabled. The post also featured a personalized message to Mark Zuckerberg of Facebook and Jack Dorsey of Twitter, mocking them.[192]

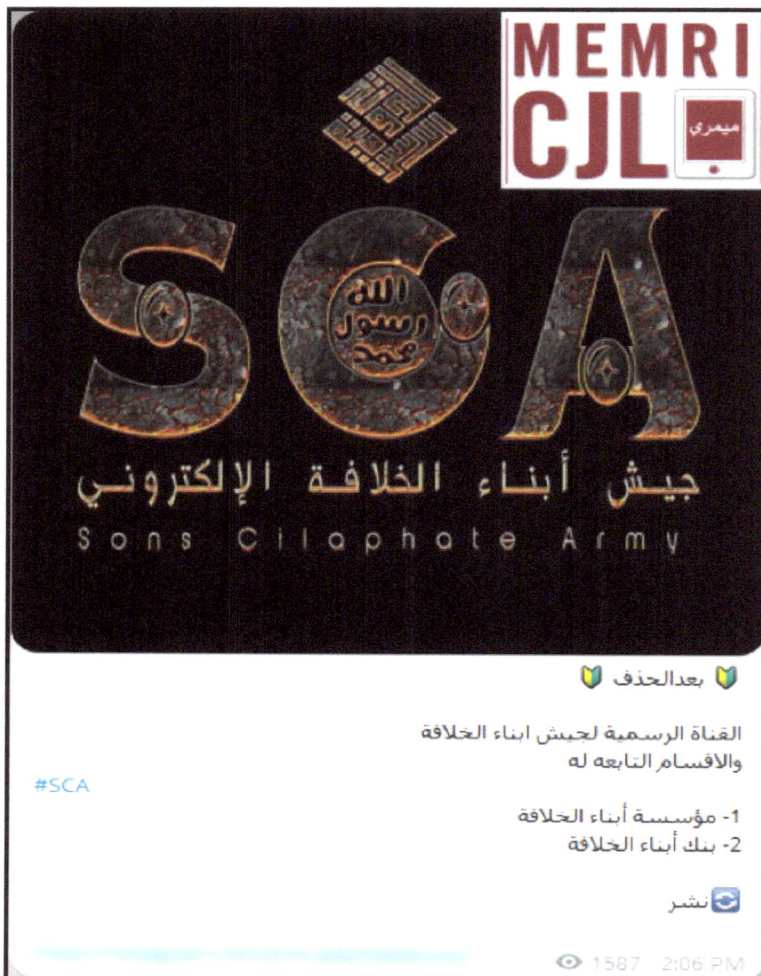

In Wake Of NY, NJ Bombings, Pro-ISIS Hacking Group Cyber Kahilafah Posts Tutorials On Making Pressure Cooker Bomb, Using Cellphones, Bluetooth For Remote Detonation, September 19, 2016

In the wake of the bombings in New York and New Jersey on September 17, 2016, the pro-ISIS hacking entity Cyber Kahilafah posted tutorials on building homemade bombs using pressure cookers, and on modifying cell phones and Bluetooth devices to denote the bombs remotely.

The tutorial was posted on the Cyber Kahilafah Telegram channel with links to the group's Dark Web website. The content included an image of the booby-trapped pressure cooker found in Manhattan. One of the videos featured surveillance footage that captured the explosion in Chelsea, NY.[193]

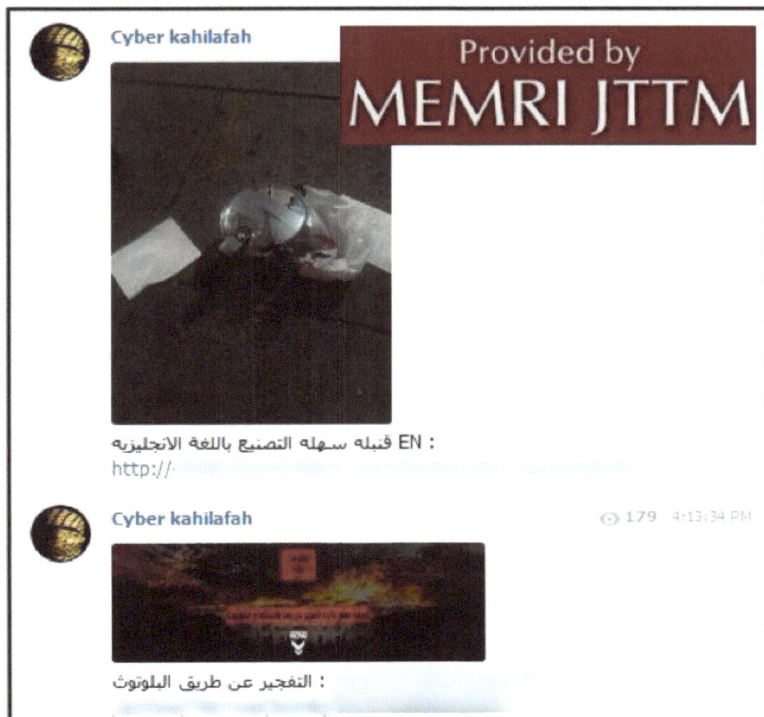

Cyber Kahilafah Informs Pro-ISIS Followers of Private Communication and Impeding Cyber Attack; Highlights Use of Online Information Sharing Platforms, April 4, 2016

On November 25, 2015, the pro-ISIS online jihadi group Cyber Kahilafah launched its official channel on Telegram. Posts on Cyber Kahilafah's channel include: pro-ISIS content, information on how to circumvent online surveillance, a link to reportedly hacked NASA data, and plans of future cyber-attacks. Cyber Kahilafah's Telegram channel also highlights Twitter's campaign to target cyber jihadi organizations, including Cyber Kahilafah itself. The channel utilizes various digital content sharing platforms including Sendvid, Google Drive, Twitter, and Justpaste.it. Cyber Kahilafah' channel also highlights the use of TOR, a software platform allowing for online anonymous communication. The variety of platforms underscores the resourcefulness of global jihadist organizations in the digital sphere. On November 27, the channel posted IP addresses belonging to "enemies of the [Islamic] State who launch spam campaigns against its supporters." Cyber Kahilafah also posted a password to access the IP list on a Ghostbin account; Ghostbin is a secured website that provides encryption and expiration of information shared online. On December 22, 2015, Cyber Kahilafah advertised that it would publish new methods for the encryption of information. A few days later, it published a list of reportedly pro-ISIS websites that was originally published on Pastebin. On January 3, 2016, it published a video with cartoon animation as well as step by step instruction on the making of an Improvised Explosive Device (IED). On February 4, 2016, it posted a link with apparently hacked NASA information; this post followed January 2016 reports that 250 gigabytes of NASA data had been hacked on behalf of the group AnonSec, and that the data included flight logs, radar logs, video taken from NASA aircraft, and employee information including names, addresses, and phone numbers. On February 16, the group promoted tools for engaging in electronic warfare.[194]

Qatar National Bank Database Hacked, Info Leaked Online Including Details On Al-Jazeera Reporters, Qatari Royal Family; Information Posted By Cyber Kahilafah, April 27, 2016

On its Telegram channel on April 26, 2016, Cyber Kahilafa posted a direct link to a leaked file of data from a hack of the Qatar National Bank database. The 1.4 GB of data included hundreds of thousands of records such as customer transaction logs, account and credit card numbers, and information on customers including several Al-Jazeera journalists and even members of Qatar's Al-Thani royal family. One folder in the leaked data is labelled "SPY, Intelligence" and contains records allegedly belonging to British MI6 and the Qatari State Security Bureau. Other folders also allegedly contain info on Polish and French intelligence.[195]

Cyber Kahilafah Posts Dark Web Advertisement For Hiring Albanian Assassins And Hackers: "You Don't Have To Pay Till The Job Is Done," June 20, 2016

On June 14, 2016, Cyber Kahilafah shared an advertisement for a Dark Web site that is allegedly run by the Albanian mafia, and offers the services of Albanian assassins and hackers.[196]

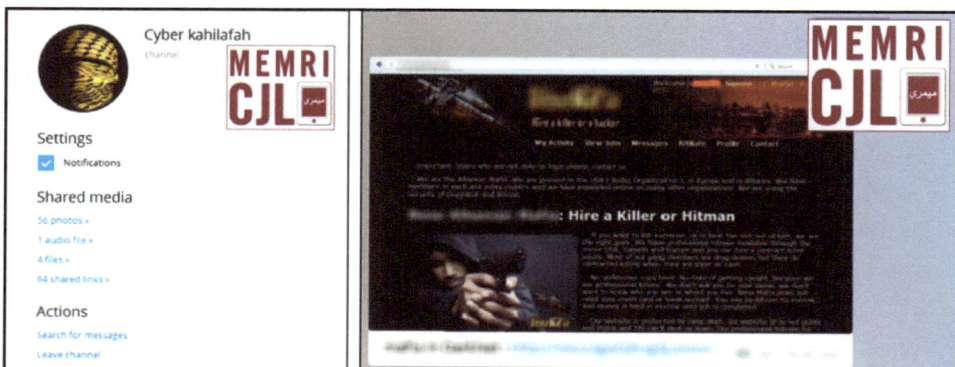

Cyber Kahilafah Telegram Channel Announces Upcoming Hack Of A SCADA System, Hinting At London, August 5, 2016

On August 5, 2016, the pro-ISIS Cyber Kahilafah Telegram channel posted a message that read "Hacking SCADA. Soon Soon." SCADA (Supervisory Control and Data Acquisition) systems are used for remote monitoring and control and are often used to control and manage utility and transportation infrastructure. The Cyber Kahilafah post included an image taken from the movie *London Has Fallen*, showing a London power-grid SCADA-type system being hacked by malicious code. This could indicate the desired target of the attack.[197]

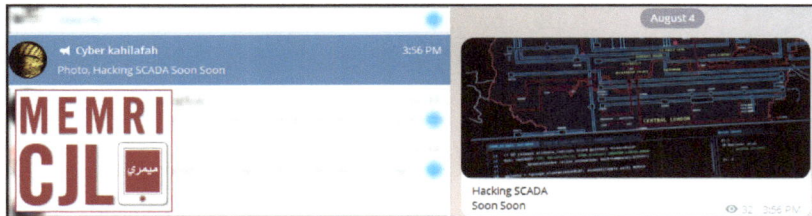

Cyber Kahilafah Announces Plans To Publish Instructions For Remote Vehicle Operation Over Wi-Fi, September 26, 2016

On September 26, 2016, Cyber Kahilafah announced on its Telegram channel that it would soon be publishing instructions on how to modify a vehicle for remote operation using Wi-Fi. The group stated that such remotely-operated vehicles could be used to spread poisons, gather intelligence, or detonated remotely. The announcement read as follows: "In the near future we intend to launch a project that will include instructions on how to modify vehicles for remote operation using a Wi-Fi network and use them for various missions such as gathering intelligence or booby-trapping and detonating them remotely. Additionally, the vehicles could be used to spread poisons while maintaining a safe distance between the target and the mujahid brother. The wireless hardware in the vehicle could also be fitted with equipment to disrupt enemy communications. While in this case we would lose contact with the vehicle, but we will still attain our goal – disrupting enemy communications. "The project is still in its infancy, but we will present it once it is finalized."[198]

Cyber Kahilafah Hacking Group Previews "Jehad Archives" On The Darknet On Telegram, September 9, 2016

In a post on Telegram on September 6, 2016, the pro-ISIS hacking group Cyber Kahilafah posted content soon to be posted on their website on the Darknet.[199]

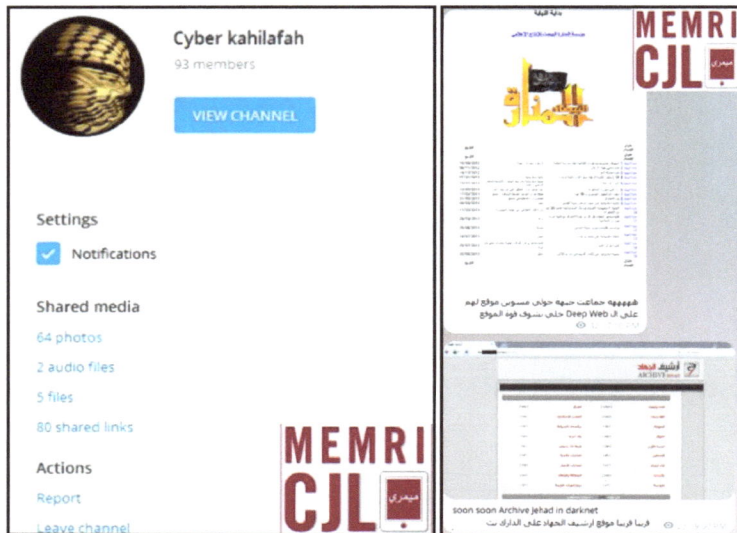

Pro-ISIS Telegram Channel Forwards UCC Instructions on VPN, April 28, 2016

On April 28, 2016, the pro-ISIS Telegram channel Online Dawah Operations forwarded a post from United Cyber Caliphate regarding the use of Virtual Private Networks (VPN). It should be mentioned that a short time previously, the channel of the pro-ISIS hacking group United Cyber Caliphate had become private.[200]

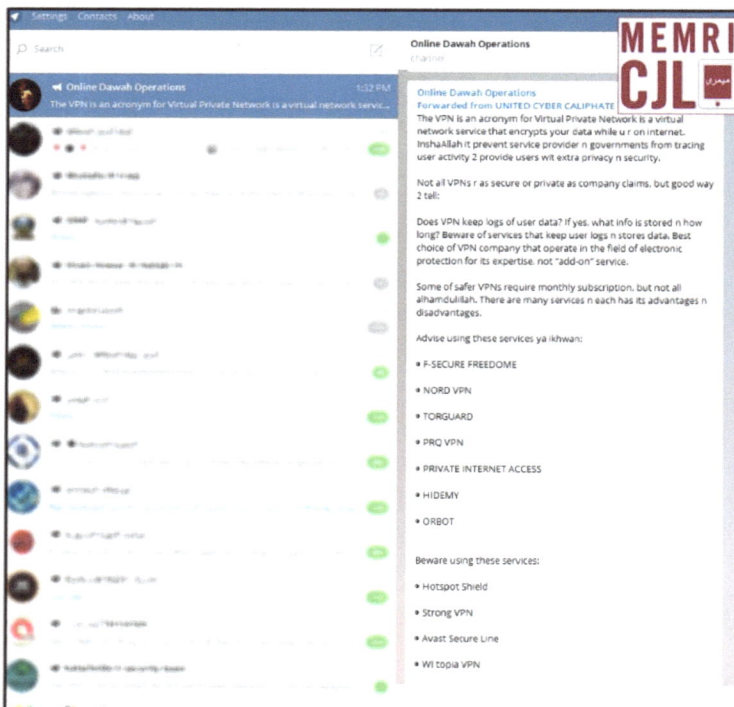

Kalachnikv E-Security Team Operates On Telegram And Other Social Media, Publishes Addresses of Federal Reserve Board of Governors, April 4, 2016

On March 3, 2016, Kalachnikv E-security team launched its official channel on Telegram. According to the group, its aim is "to publish various web-hacking techniques, exploits, [and] articles with... information and [the] latest cyber attacks." On March 6, 2016, another Telegram channel was launched under the name Kalachnikv Team. Both channels are primarily in English and target a Western audience, but both frequently cite "Mosul time," indicating possible geographic ties with the Iraqi city, which is under ISIS control. The Kalachnikv E-security team channel places greater emphasis on technical support to jihadi cyber activists, while the Kalachnikv Team channel acts as a mouthpiece, uploading ISIS-related jihadi literature, sharing posts from cyber jihadi groups, and reporting successful attacks on websites and Facebook pages. The separation of Kalachnikv's technical and propaganda activities underscores its aim to boost cyber jihadi enthusiasm while providing technical education to current and potential team members.

On March 4, 2016, Kalachnikv E-security team boasted of an attack on an Italian web server, claiming it had gained "full access." It also provided a description of what a web server is for the potential hacker. On March 8, it posted a list of websites it had allegedly hacked. On March 9, it published the alleged addresses of current and former members of the United States Federal Reserve Board of Governors, including Ben Bernanke and Janet Yellen. On March 17, it highlighted the risk of followers being located while using electronic devices and related services. The group also warns followers using mobile devices that "some websites," regardless of whether or not the user consents to the use of geolocation, have locating capabilities, and advises followers to use prepaid phones as a method to mitigate the aforementioned concerns. Following the March 22, 2016 ISIS attacks in Brussels, Kalachnikv E-security team forwarded security measures directed to "brothers in Belgium."[201]

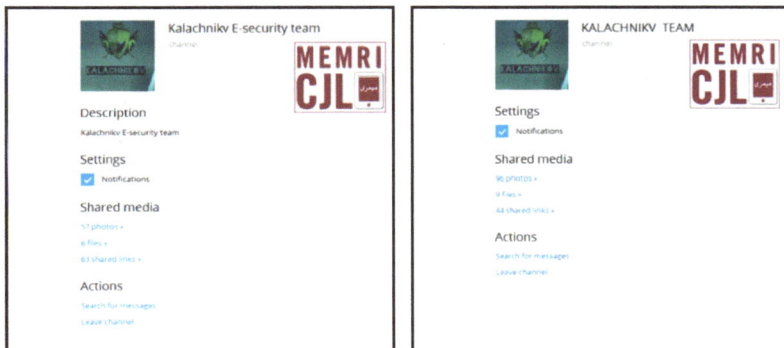

On Telegram, Kalachnikv E-security Team Posts Image Threatening Mark Zuckerberg, March 21, 2016

On March 17, 2016, the pro-ISIS hacking group Kalachnikv E-security team uploaded an image, on its Telegram channel, of Facebook cofounder Mark Zuckerberg facing a noose. The group frequently posts articles on web hacking techniques along with encryption tips.[202]

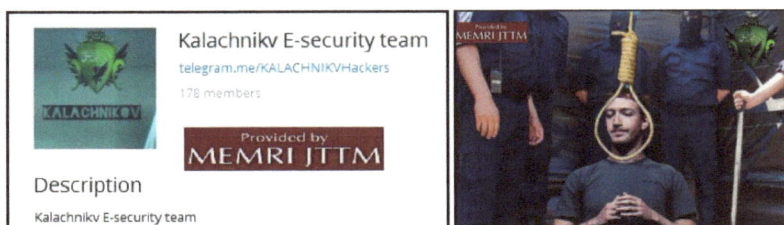

In Kill List On Telegram, Kalachnikv E-security Team Posts Names Of "Most Important Crusaders In Texas," May 3, 2016

On May 3, 2016 the pro-ISIS hacking group Kalachnikv E-security team, an affiliate of the United Cyber Caliphate, shared a graphic on Telegram that reads: "List contain most important crusaders in Texas. Wanted to be killed. 'Crush the cross' SHOOT THEM DOWN. #Islamic State" along with a list of over 1,000 individuals and their personal information. MEMRI has the full list of names. The personal information includes home addresses, emails, and phone numbers. Over the past couple of weeks, the same hacking groups have released the personal details of U.S. State Department employees as well as of New York City and other U.S. residents.

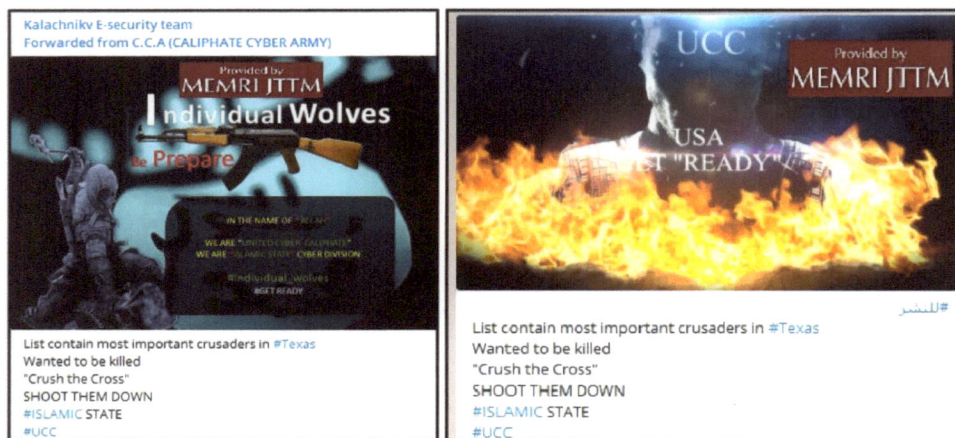

Kalachnikv E-Security Team Announces New Member From TheDarknet Nation, Provides Info On Various Security Software, May 4, 2016

On May 3, 2016, Kalachnikv E-Security Team announced that a member of "TheDarkNet Nation" named 3aox had decided to join the group. The user provided tips and tricks on security including how to defend oneself against various forms of hacking and phishing.[203]

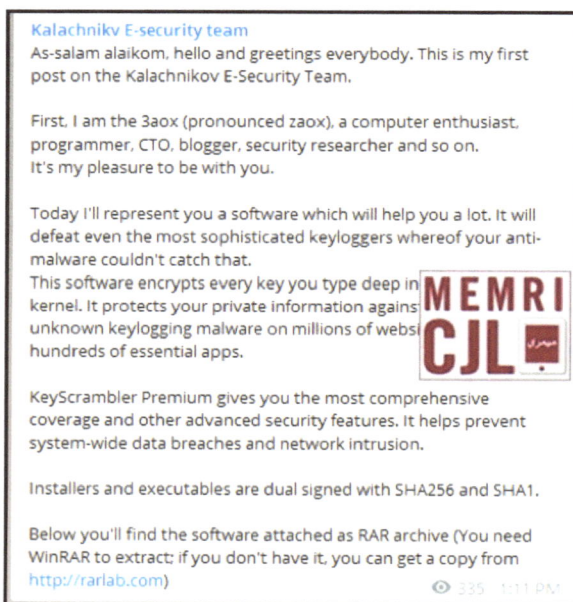

Kalachnikv E-Security Team Participates In #OpSaudiArabia, May 11, 2016

On May 8, 2016, Kalachnikv E-security Team announced the launch of #OpSaudiArabia – a joint cyber-attack on Saudi websites in response for "killing kids and civilians in Yemen with Government air strike." The group announced the Op on their Telegram channel.[204]

Kalachinkv E-Security Team Hacks Indonesian Site, Threatens U.S. Government, May 25, 2016

On May 25, 2016, Kalachnikv E-security Team announced on its Telegram channel that it had hacked the "military leadership in Pariaman" – Pariaman is a coastal city on the Indonesian island of Sumatra. The group also threatened the U.S. government, saying: "We are willing to attack American government sites soon! This time it'll be a bigger and better breach."[205]

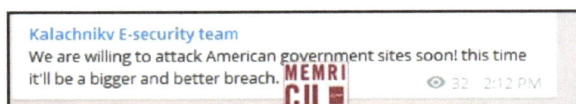

Caliphate Electronic Army Recruiting New Members, May 4, 2016

On April 29, 2016, pro-ISIS American jihadi Telegram user Amriki Muhajer shared a post about recruitment efforts for the hacking group Caliphate Electronic Army. The post read: "Important announcement! The recruitment office is looking for new recruits who are interested to join CEA brigades to work and support their Muslim brothers. To register, please write to brother War Minister."[206]

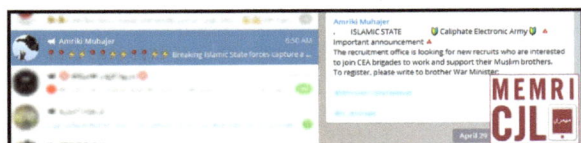

Pro-ISIS Hacking And Tech Group Rabitat Al-Ansar Shares Tech-Related Information On Telegram, June 20, 2016

Rabitat Al-Ansar, a pro-ISIS hacking collective, is active on Telegram. The group's account, launched on March 25, 2016, provides ISIS sympathizers with instructions on how to download software and programs to increase their online privacy and security. Rabitat Al-Ansar is a part of a larger network of media, tech and hacking entities supporting ISIS. The group also publishes various tech-related information, such as instructions on using different programs to increase users' safety when using Android devices, and on using VPNs for security and anonymity. Many of the phone-related instructions pertain to Android devices.

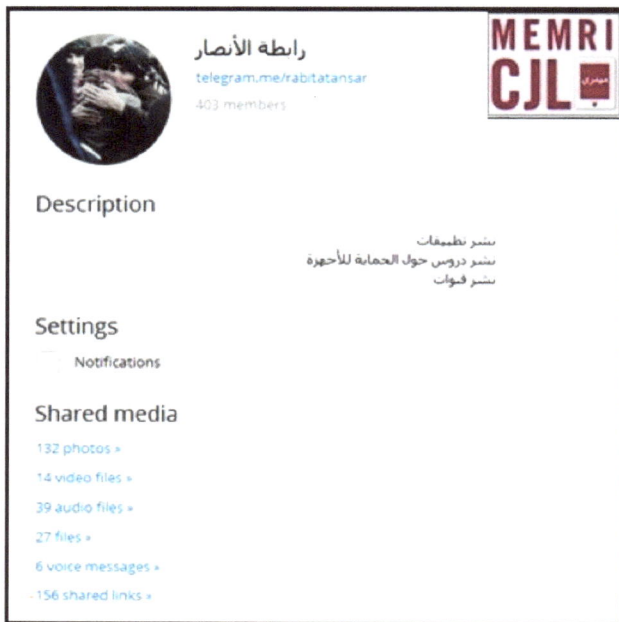

Pro-ISIS Hacking Group Cyber TeamRox (CTR) Active On Telegram, March 9, 2016

On March 3, 2016, the pro-ISIS hacking group Cyber TeamRox (CTR) launched its official channel on the encrypted messaging app Telegram. Posts by the group include: hacking technique; articles about cyberattacks and contributions from members; instructions on using VPN, how to circumvent online surveillance, and hacking drones; tracking the U.S. government's cyber efforts, and claims of hacking the NSA; monitoring the Pentagon and other U.S. agencies; and recommendations to follow other jihadi and pro-ISIS accounts, including the "Constants of Jihad."

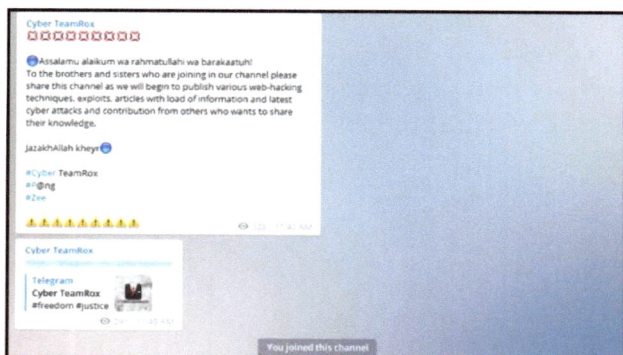

The group claims to have hacked the NSA and Saudi government websites.[207] On March 3, the group posted a message setting out its goals: publishing various hacking techniques and exploits, informative articles, and announcements of the latest cyber attacks, as well as contributions from others who want to share their knowledge. It published a list of "do's and don'ts" in hacking to ensure that actions do not draw attention or surveillance. Some of the tips focused on how to avoid getting on watchlists and how to circumvent surveillance, and warnings to not deal with the FBI. It published what it said was the details of a hack of the NSA, forwarded from the "Daily hadith" Telegram channel, and also posted a link to instructions on drone hacking:

Pro-ISIS Tech Channel Warns Of Alleged ISIS Telegram Groups Offering Hacking Tutorials, May 19, 2016

On May 17, 2016, the Islamic State Technician Telegram channel, which is pro-ISIS, posted a warning to Telegram users to beware of entities on the platform offering hacking lessons while claiming official affiliation with the Islamic State. The announcement read: "Important warning: some groups on Telegram have publicized a project for training hackers among the [Islamic State] supporters, and [claiming] that they [i.e. the groups] are officially connected with the Islamic State. And based upon that, we warn the brothers and sisters of their lies, and point out that no one represents the Islamic State on the communications networks but [its] official media companies. And we warn of any technical projects that claim official connection with the Islamic State..." The Islamic State Technician channel is part of a larger nexus that caters to jihadis' technical needs.[208]

Islamic Cyber Army Issues Security Directives On Telegram Following Shutdown Of Twitter Account, December 8, 2015

On December 8, 2015, the ISIS-affiliated Islamic Cyber Army issued a message via Telegram on the shutdown of the Cyber-Caliphate Twitter account. The message included security directives to the group's followers stating claiming that a virus was attached to a kill list CyberCaliphate had posted on Twitter.[209]

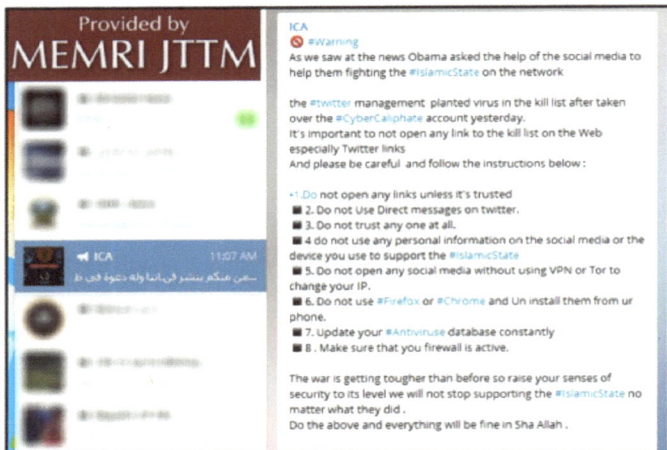

Designated Terror Group Ansar Al-Shari'a In Libya Begins Using Telegram To Disseminate Its News, May 6, 2015

On May 2, 2015, the Atheer Al-Madina radio channel, which belongs to the militant group Ansar Al-Shari'a in Libya (ASL), announced that it will begin sending its news via the instant messenger app Telegram.[210]

Al-Qaeda: Al-Qaeda Central; Al-Qaeda Media Organizations; Al-Qaeda In The Arabian Peninsula (AQAP); Al-Qaeda In The Islamic Maghreb (AQIM); Al-Shabab Al-Mujahideen

Al-Qaeda Central

Al-Qaeda Leader Ayman Al-Zawahiri Calls On Iraqi Muslims To Unite, Wage Guerilla Campaign To Rid Iraq Of Shi'ite-Crusader Occupation, August 25, 2016

On August 25, 2016, the official Al-Qaeda media body Al-Sahab published, inter alia via its Telegram channel, a four-minute video in which the group's leader, Ayman Al-Zawahiri, called on Sunni Muslims in Iraq to unite their ranks and launch a lengthy guerilla campaign to expel the Shi'ite-Crusader occupation from their country. According to Al-Zawahiri, an Iranian-Shi'ite-Crusader campaign is currently underway against Iraq's Sunnis on the pretext of fighting "the Ibrahim Al-Badri group" – a pejorative reference to ISIS and its leader Abu Bakr Al-Baghdadi. However, Al-Zawahiri alleges that the true goal of this campaign is to exterminate the Sunni population. He argued that the events in Syria were not merely a localized problem, but rather a tragedy for all Muslims. He called on the mujahideen in Syria to assist their comrades in Iraq and reorganize, as their campaign is a joint one. This is the third video in the "Brief Messages to a Victorious Nation" series, in which Al-Zawahiri delivers various messages to the Muslim ummah.[211]

Al-Zawahiri Calls To Unite Jihadi Groups, Accuses ISIS Of Causing Schism And Harming Jihad, September 1, 2016

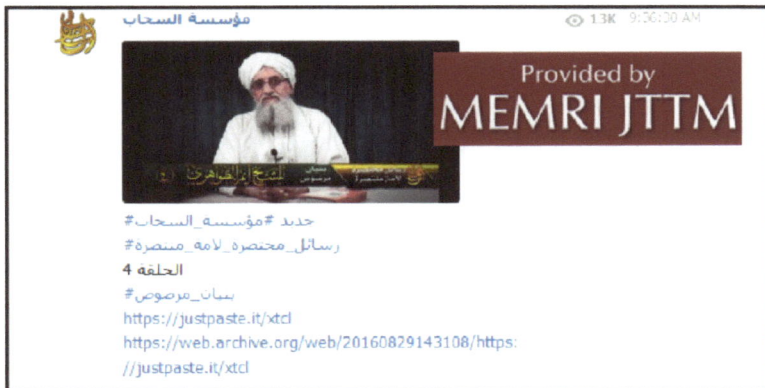

Days after the release of a video calling on Sunni Muslims in Iraq to unite and expel the Shi'ite-Crusader occupation, on August 30, 2016, Al-Qaeda's official media wing Al-Sahab released a 15-minute video titled "One United Structure" featuring Al-Zawahiri. In the video, which was also posted on the group's official Telegram channel, Al-Zawahiri called on all mujahideen everywhere to join ranks under one united body to concentrate their efforts fighting the enemies of Islam. Al-Zawahiri repeatedly attacks "Ibrahim Al-Badri and his gang" – a pejorative reference to Abu Bakr Al-Baghdadi and ISIS – claiming that they have divided the ranks and caused harm to the path of jihad.[212]

Al-Qaeda Gives Jabhat Al-Nusra Green Light To Leave Its Fold, July 28, 2016

On July 28, 2016, Jabhat Al-Nusra (JN) released, via its official Telegram channel, an audio message from Al-Qaeda deputy leader Ahmad Hassan Abu Al-Khair Al-Masri, announced that following consultation among its senior leadership, Al-Qaeda had decided to instruct JN to "pursue whatever measures" are necessary to protect the jihad in Syria and unite the jihad factions in Syria. By this, Al-Qaeda has effectively complied with JN's request to sever its ties with Al-Qaeda, its parent organization, so as to promote unity among the Islamist militias in Syria. Following general remarks about the situation in the Muslim world, Abu Al-Khair Al-Masri said in the message: "We are living in a blessed, honorable era... an era of awakening and jihad. People have risen up against their tormentors. They have launched jihad with words and weapons..." The message is followed by an excerpt from an old recording by Al-Zawahiri stating that Islam unites all Muslims and jihad groups, and that unity among Syrian factions is more important than all organizational ties.3[215]

Al-Qaeda Media Organizations – GIMF And Al-Sahab

GIMF Announces Its New Channel On Telegram, October 6, 2015

On October 5, 2015, the Al-Qaeda-affiliated media company Global Islamic Media Front (GIMF) announced its new Telegram channel.[214] GIMF is one of Al-Qaeda's most important media outlets.

As of this date, GIMF was also using Telegram's private chat feature.

GIMF Creates Telegram Channel For Jihadi Content Related To Indian Subcontinent, August 9, 2016

On August 4, 2016, the GIMF announced on its Telegram channel that it had launched a special channel, titled GIMF Subcontinent, which will focus on disseminating content pertaining to the Indian subcontinent. GIMF attributed the move to the need to increase its outreach (da'wa) efforts to people in that region, as well as the increased influx of jihadi-related content pertaining to that region.[215]

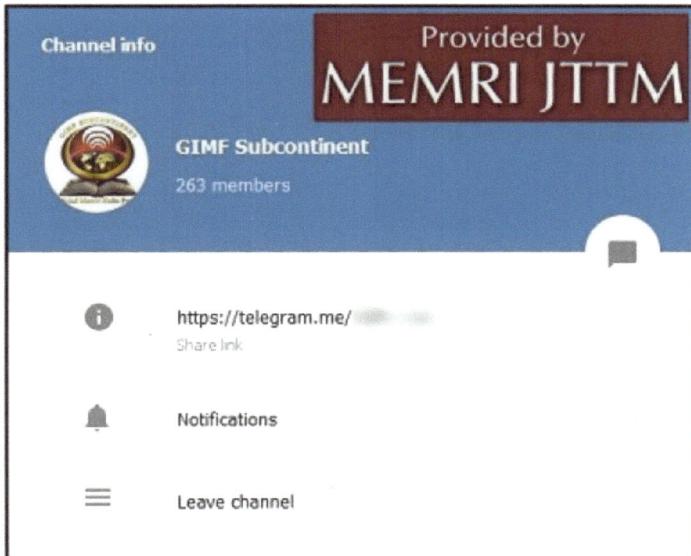

On Telegram, Global Islamic Media Front (GIMF) Shares Technical Advice For Aspiring Jihadis, February 24, 2016

The technical section of the Al-Qaeda-affiliated GIMF operates a Telegram channel for posting updates and news-related items on variety of cybersecurity topics.[216]

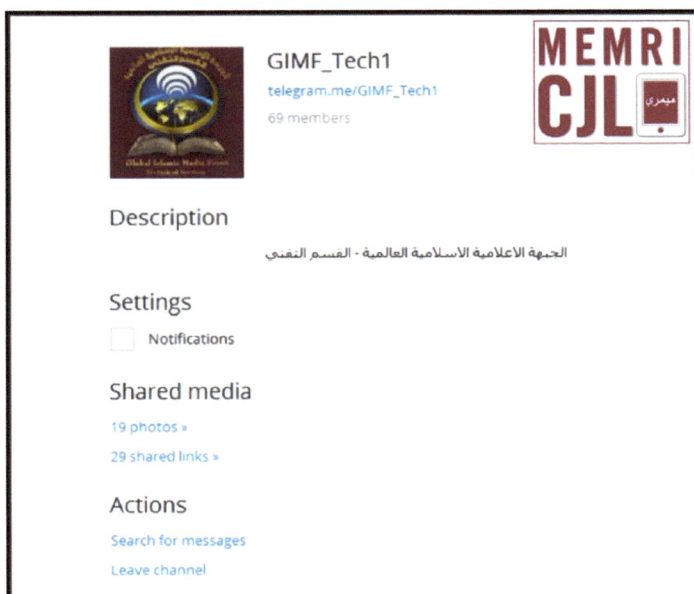

Al-Qaeda Media Wing Al-Sahab Active On Telegram, August 31, 2016

The Al-Qaeda media wing Al-Sahab's Telegram channel was created on July 3, 2016 and as of August 29, 2016, had 3,139 members and posted 65 items. While most of the posts are in Arabic, there are some in English, mostly "Al-Nafir" news bulletins translated from Arabic. Some of the posts link to videos featuring Al-Qaeda leader Ayman al-Zawahiri. The most recent post features a video of Zawahiri, along with English-language links to content on JustPaste.it and the Internet Archive.[217]

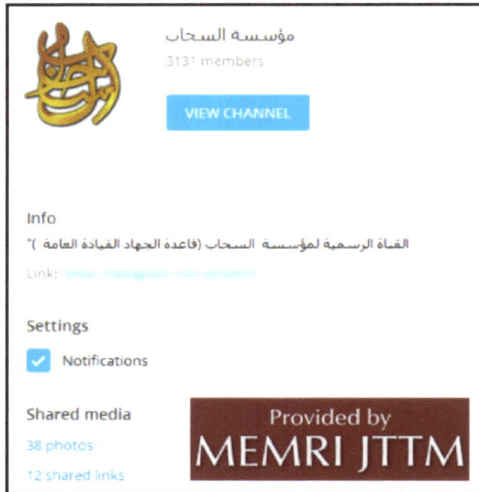

Al-Qaeda Telegram Account Promotes 9/11 Anniversary Video, September 6, 2016

On September 6, 2016, the Telegram channel of the Al-Qaeda media wing Al-Sahab promoted an upcoming video release for the 15th anniversary of the 9/11 attacks, titled "Rejecters of Tyranny."[218]

Al-Qaeda In The Arabian Peninsula (AQAP)

AQAP Weekly Magazine Calls Hillary Clinton, Donald Trump "Full Of Flaws," Says It Is Time For U.S. Prominence To Decline, October 26, 2016

On October 26, 2016, *Al-Masra*, the weekly newspaper of Al-Qaeda in the Arabian Peninsula (AQAP), published an article describing U.S. presidential candidates Hillary Clinton and Donald Trump as "full of flaws," stating that the future of the U.S. will be different from its shining past, and that it is time for U.S. prominence to decline. The article, by Ibrahim Abu Al-Futouh, mentioned the recriminations between the two candidates, as well as Trump's implication that he may not concede the results if he loses, and the alleged Russian involvement in leaking Clinton's emails in order to tarnish her credibility and boost her opponent's chances in the election. According to the article, Trump has succeeded in shaking people's trust in the government, the American people are extremely divided internally, and even if Clinton won, she would be unable to reunite the country. "The U.S.," the article noted, "is no longer a united front in Washington that is ready to face growing challenges." After describing Clinton and Trump as "full of flaws," the article stated: "Is this an indication of the beginning of the collapse of the U.S.? The answer will be confirmed by the results of the election. Undoubtedly, the future will be different from the past history, during which the star of the U.S. shone, and it is time for it to decline."[219]

Telegram Channel Disseminates Works Of Al-Awlaki, December 15, 2015

On December 11, 2015, a Telegram channel called "Imam Anwar Al Awlaki rh" was created; it disseminates the works of the radical Yemeni-American sheikh. It shares audio files from Awlaki's seminal "Hereafter" series, along with a number of video interviews with him. The channel's information page directs followers to Kalamullah.com for additional audio recordings, videos, and books by Al-Awlaki. As of this date, the Telegram channel had 423 members.

One post on the channel explains that expanding this Telegram channel is crucial, because much of Al-Awlaki's material have been removed from the Internet: "Dear Brothers and Sisters Asalamaualaikum. The enemies of Allah was so scared of Sh. Anwar Al Awlaki rh and his message which penetrated the hearts of thousands of Muslims, so they killed him. Now by the grace of Allah, all of his works has been with many of the Muslims worldwide even though Disbelievers tried to delete and remove most of his works from the internet. So we consider is our duty to spread his teaching and books etc. We ask all who love the Sheikh to spread this channel and earn the ajar [reward] spreading knowledge."[220]

Jihadis Promote New AQAP Channel On Telegram, September 29, 2015

On September 26, 2015, a jihadi Twitter account reported that Al-Qaeda in the Arabian Peninsula (AQAP) now has a channel on Telegram. The channel, according to the tweet, will be used to disseminate the group's news and releases. [221]

Issue 20 Of AQAP Weekly *Al-Masra* Announces Launch Of Official Telegram Channel, September 6, 2016

On September 3, 2016, Al-Qaeda in the Arabian Peninsula (AQAP) released Issue 20 of its weekly *Al-Masra*. This issue included the announcement that the Telegram channel UmmahNewsCH3 is the official and exclusive distributor of this publication. [222]

Telegram Channel Devoted To AQAP's *Inspire* Magazine Posts Bomb-Making Instructions, March 18, 2016

A Telegram channel called Inspire Muslims shared, on March 17, 2016, pages from the well-known article from the first issue, published in the summer of 2010, of the AQAP English-language magazine *Inspire*, "Make a bomb in the kitchen of your Mom." The shared pages included step-by-step instructions and images on bomb-making.

One message in the channel reads, "Simple and easy to make above and give kuffar [infidels] some special gifts on special occasions on special places."[223]

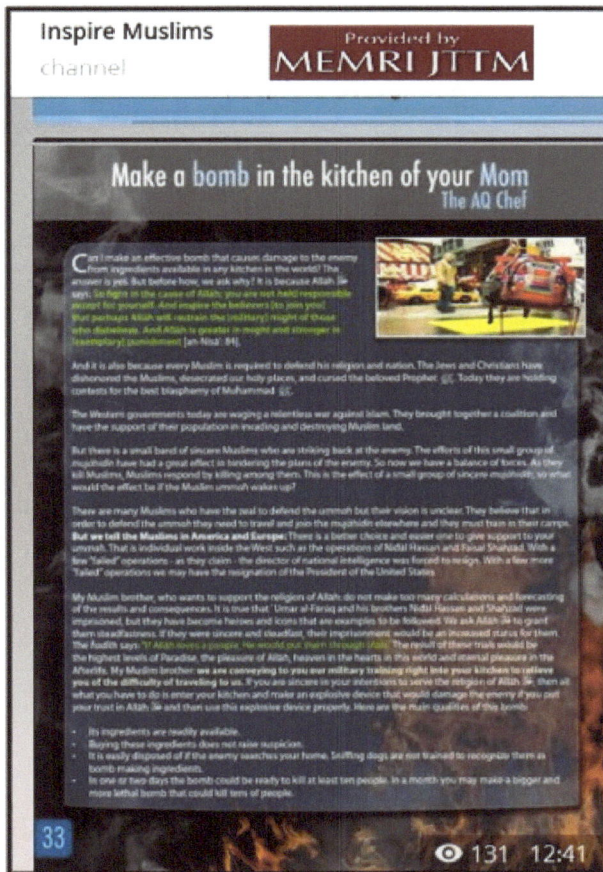

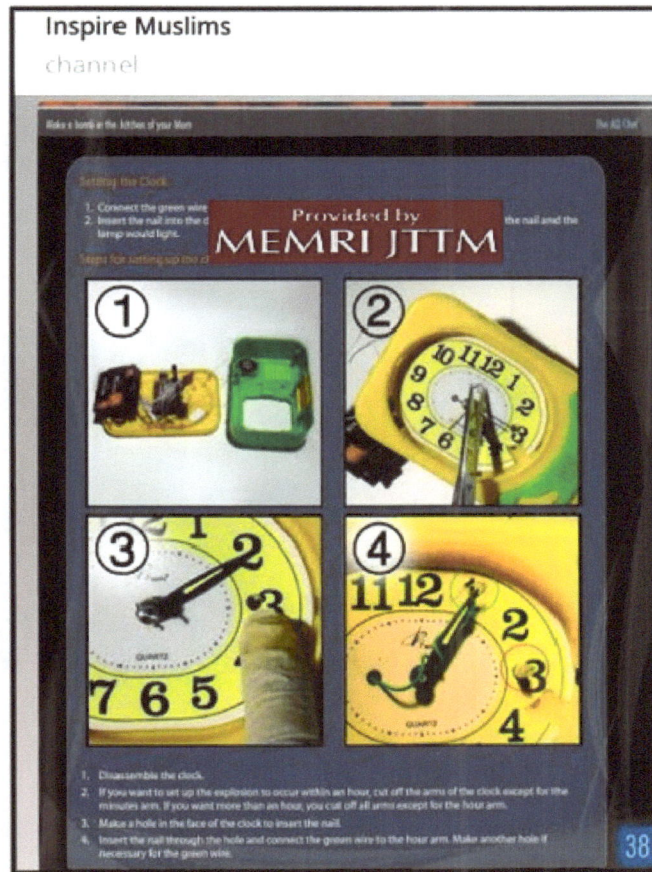

In Video, AQAP Presents Special Operations Brigade, Vows To Fight Until Liberation Of Al-Aqsa Mosque, July 13, 2016

On July 13, 2016, Ansar Al-Sharia Correspondent and Al-Malahem, the media branches of Al-Qaeda in the Arabian Peninsula (AQAP), released a video presenting members of the group's Special Operations Brigade demonstrating their skills at a training camp named for Hamza Al-Zinjbari aka Jalal Bal'idi aka Hamza Al-Marqashi – a senior commander who was killed in February 2016. The video, titled "Commander Hamza Al-Zinjbari Training Camp" and posted on the group's official Telegram channel, featured Khaled Batarfi and Ibrahim Al-Qusi, both top AQAP commanders, speaking of the importance that Muslims be prepared physically, mentally, and financially for jihad, and vowing that AQAP will continue fighting until Al-Aqsa Mosque in Jerusalem is liberated.

Speaking to those who have already been trained but are neglecting their training, Batarfi noted that this situation is strongly denounced in Islam, and added: "Then how about the ones who have never been prepared [militarily] at all and who have never cared about this obligation [jihad] although it is compulsory and [highly] important at this time, when the nations of unbelief have united against us and when the Crusaders and Jews and their supporters have marched forth... to eradicate and exterminate us? Where are the free and righteous men among the youth of the *ummah* to respond to the calls for [military] readiness and for jihad to terrorize the enemy of Allah?" Praising the camp's namesake Al-Zinjbari, Al-Qusi said: "This is the special operations training camp [named for Al-Zinjbari], and Hamza was also the leader of special operations and not only a military commander." Addressing the U.S., Al-Qusi downplayed the significance of Al-Zinibari's death, saying that the U.S. is still far from achieving its goal as "Hamza's students are in the field holding the banner and have sworn that they will continue on the same path. The images speak louder than words."[224]

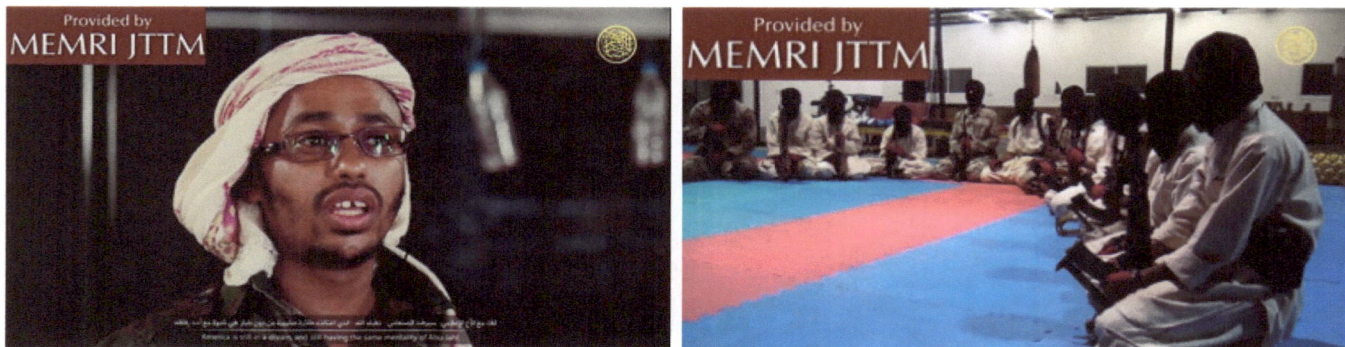

Telegram Channel Distributes Issues Of AQAP *Inspire* Magazine, March 9, 2016

On February 22, 2016, a Telegram channel called "Inspire Magazine" was created to disseminate issues of the English-language Al-Qaeda in the Arabian Peninsula (AQAP) magazine *Inspire*. Users can download PDF copies of issues of the magazine via the channel. The first post in the channel states: "Inspire 1 This is the first edition of Inspire magazine which inspired hundreds of Muslims throughout the Globe. This is from the Land of Iman [belief] and Wisdom-Yemen. Till this date 14 issues were published out of which 12 were uploaded in this channel. Remaining is only Inspire 2 after this one. InshAllah soon willl be uploaded."[225]

An item posted in the channel, after a link to Issue 13 of *Inspire* was posted, reads: "Inspire 13 was reportedly downloaded 50000 times using UK IP address. My brothers and sisters, Spread the magazine to your beloved ones and enrage the Kuffar [infidels]."

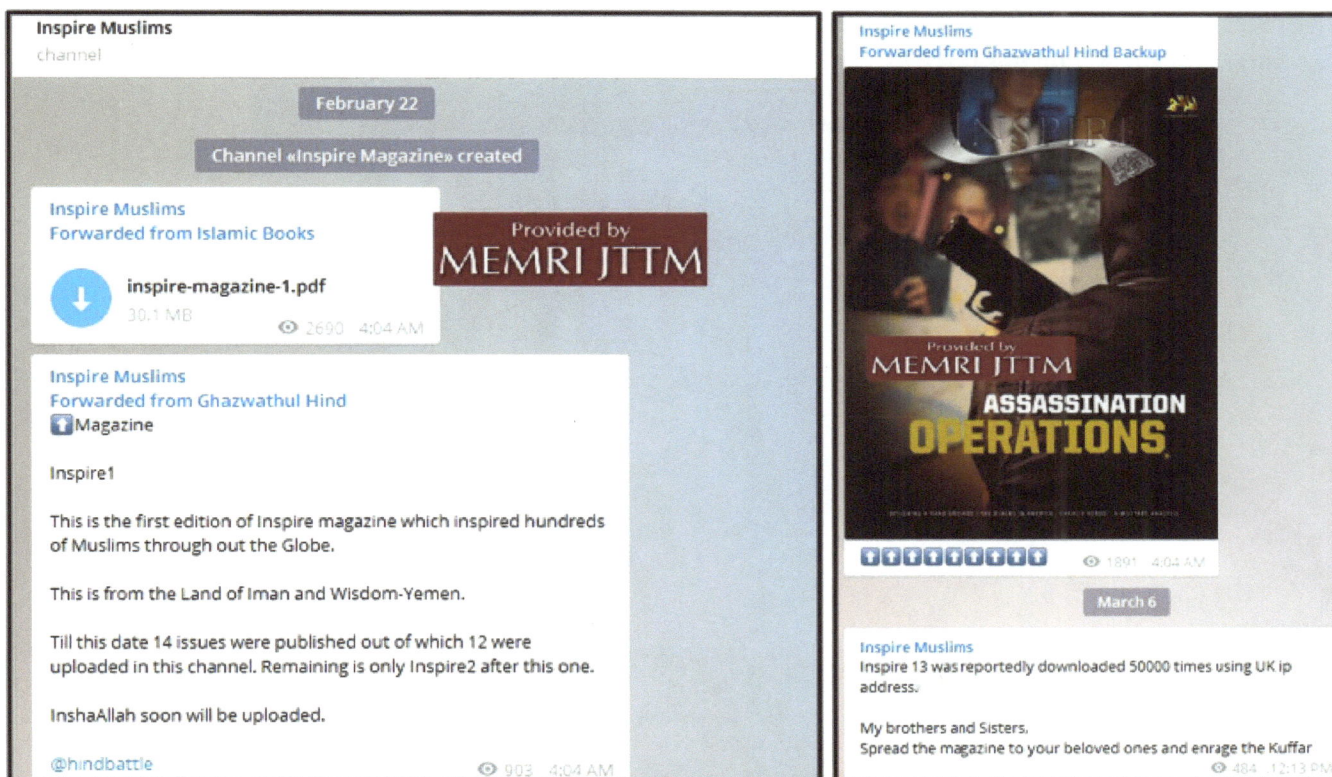

Pro-AQAP Telegram Channel Provides Instructions For Lone Wolf Attacks At Rio Summer Olympics; American, French, Israeli, British Athletes Singled Out, July 19, 2016

On July 19, 2016, a pro-Al-Qaeda in the Arabian Peninsula (AQAP) Telegram channel titled "Inspire the Believers!" posted a list of 17 suggestions for lone wolf attacks during the 2016 Summer Olympics in Rio de Janeiro, Brazil, and included an English-language schedule of events for the Games. The Telegram post encouraged lone wolves by claiming that travel to Brazil is relatively cheap and easy: "Lone wolf [sic.] from anywhere in the world can move to Brazil now. Visas and tickets and travel to Brazil will be very easy to get inshaAllah." Suggestions for attacks include attaching small explosives to toy drones, perpetrating a knife attack against Americans and Israelis, and entering bars and pubs in the area to attack, kidnap, or rob drunk patrons. The 17 points included: "Olympics will start within 17 days insha Allah. This time its [sic.] in Brazil. Which means we have targets from all the countries in war with us representing their countries there! (For schedule and more info of the Olympics. Check this link https:// www.rio2016.com/en)"; "Lone wolf [sic.] from anywhere in the world can move to Brazil now. Visas and tickets and travel to Brazil will be very easy to get insha Allah."; "The Combatant enemy (Your target) will be coming there insha Allah... Insha Allah your chance to take part in the global Jihad is here! Your chance to be a martyr is here! Do dua [pray] that Allah accepts from all of us!"; "Many Multinational Companies who are American/Western gain a lot [of funds] from these Olympics, which they pump into the enemy economy again fighting against the Muslims..."; "So we suggest that we Mujahideen take part in the Olympics too... How?"[226]

In Wake Of Nice Attack, Telegram Channel Devoted To AQAP's *Inspire* Magazine Reposts Images Calling For Lone-Wolf Vehicular Attacks, July 15, 2016

Following the truck attack in Nice, France on July 14, 2016, the Telegram channel Inspire Muslims, which is dedicated to AQAP's English-language magazine *Inspire*, reposted images from the magazine's second issue, which call for vehicular attack. The post reads "These kinds of Mow Attacks by big pickups were suggested by Inspire Magazine Years Ago!" The images include one of a pickup truck with the caption "The Ultimate Mowing Machine" as well as quotes from the magazine's "Open Source Jihad" section, which recommends methods for lone wolf attacks.[227]

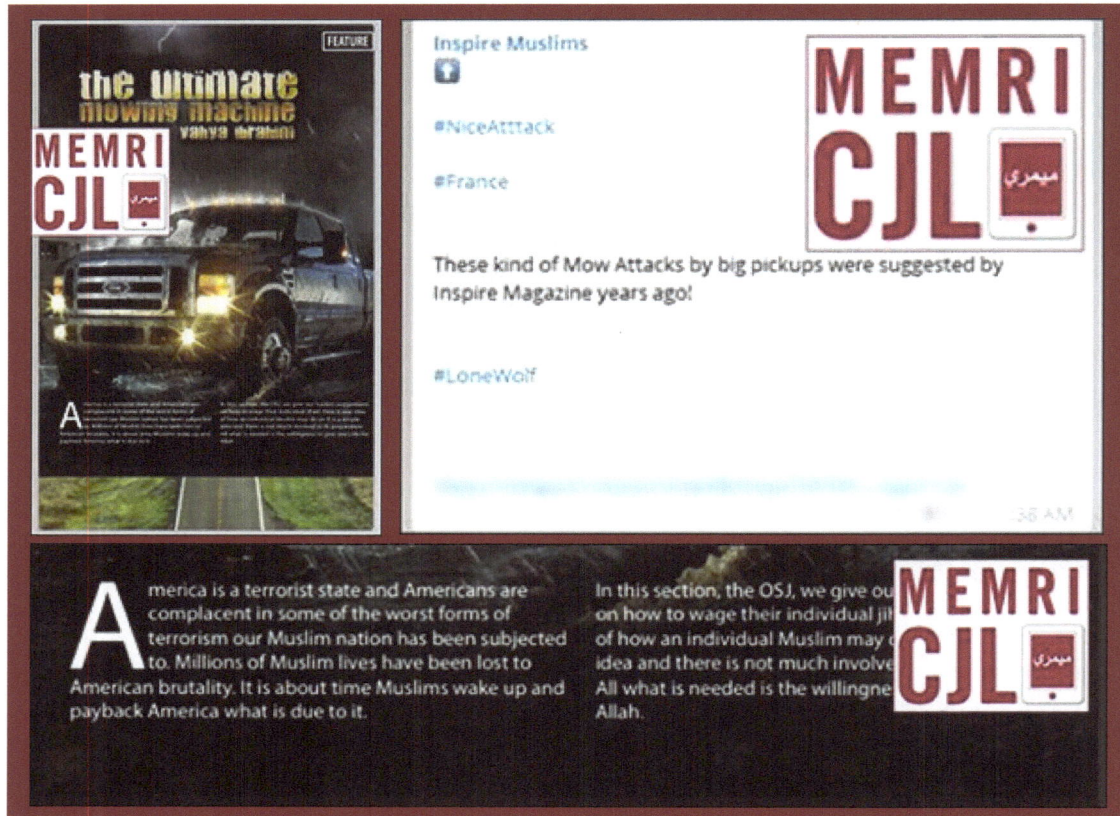

Prominent Twitter Account Documenting "U.S. Crimes In Yemen" Switches To Telegram, Expresses Frustration At Twitter's Repeated Suspensions, November 5, 2015

On November 5, 2015, a jihadi Twitter account announced the launch of a new channel on Telegram for reporting "U.S. crimes in Yemen." The switch to Telegram follows repeated shutdowns of similarly themed accounts on Twitter reporting on U.S. military operations and their repercussions in Yemen. Those behind the new Telegram channel expressed their frustration at Twitter's shutdowns, accusing it of attempting to gag an "unbiased media" source.[228]

Al-Qaeda In The Islamic Maghreb (AQIM)

AQIM Announces On Telegram The Kidnapping Of Swiss Citizen In Mali, Issues Demands, January 27, 2016

Al-Qaeda in the Islamic Maghreb (AQIM) announced in a January 26, 2016 video that it had kidnapped a Swiss national, Beatrice Stockly, and issued demands for her release. The main speaker in the 8-minute video, which was circulated via AQIM Telegram channels and Twitter accounts, is an AQIM fighter who speaks in English and claims that Stockly, "a Swiss nun who declared war on Islam," was captured by fighters from "AQIM's Sahara region" in Timbuktu in 2012, but was released on the order of its leader, and was subsequently recaptured. The video ends with a statement from Stockly herself, who says she was captured after AQIM's January 15 attack in Burkina Faso and admits to engaging in missionary work.[229]

AQIM Launches Campaign In Support Of Al-Qaeda-Affiliated Militias In Benghazi, June 28, 2016

On June 26, 2016, Al-Qaeda in the Islamic Maghreb (AQIM) launched a media campaign calling for support for Ansar Al-Shari'a and AQIM-affiliated militias in Benghazi, Libya. The campaign includes a series of images, audio messages by AQIM Shura Council member Abu Obeida Yussuf Al-'Anabi and Jabhat Al-Nusra cleric Sheikh Abdullah Al-Muhaysini, and a written statement by an Al-Qaeda fighter in Benghazi. The support campaign comes following heavy fighting inside Benghazi between Al-Qaeda militias and forces loyal to the Tobruk government. AQIM's support campaign was announced in a June 26, 2016 post on the AQIM Telegram channel Ifriqiya Al-Muslima. The post states that "Western and pro-Western Arab media are intentionally silent with regard to coverage on the total destruction of Benghazi city at the hands of the French crusaders and their puppet *taghout* Haftar, their allies in the UAE, and Egyptian dictator Al-Sisi. Our Islamic obligation toward the Muslims inside the city requires us to stand up in defense of our brothers in Libya. Therefore, we announced the start of a support campaign for our brothers in Libya..."

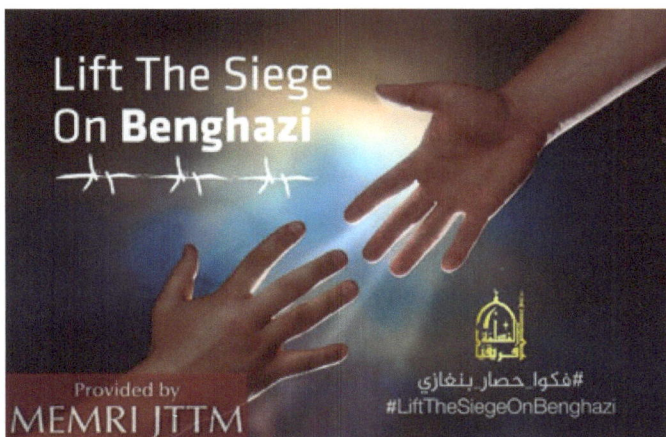
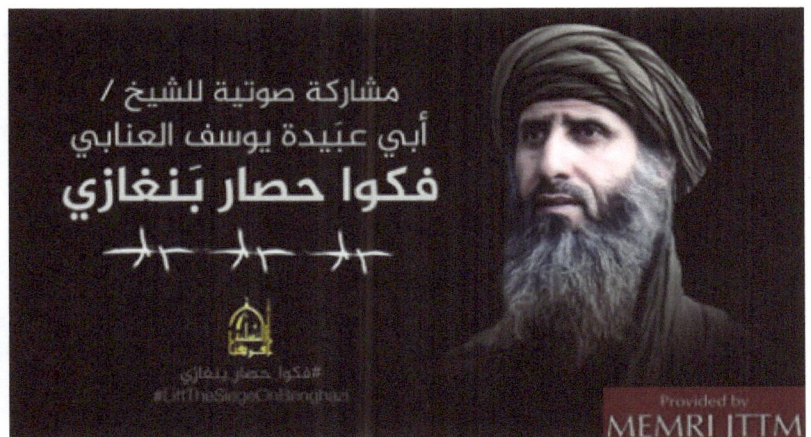

AQIM Statement: Ivory Coast Attack Was Response To France's Military Campaign In Sahel – And Vengeance Against The Countries That Participate In French Military Operations In The Region, March 15, 2016

On March 13, 2016, Al-Qaeda in the Islamic Maghreb (AQIM) released a full statement concerning its attack that day on the resort town of Grand-Bassam, Ivory Coast. Immediately following the attack, AQIM had released a brief statement claiming responsibility for it, via its Telegram channel. In the full statement, AQIM said that the attack was part of its ongoing campaign to target the Crusaders' "dens" in the region, and underlined that the resort town was in fact a "spying and conspiracy den" used by the "heads of crime and theft." Naming the three attackers as Hamza Al-Fulani, 'Abd Al-Rahman Al-Fulani, and Abu Adam Al-Ansari, the group said that these were the ones chosen to carry out "this grand battle."[230]

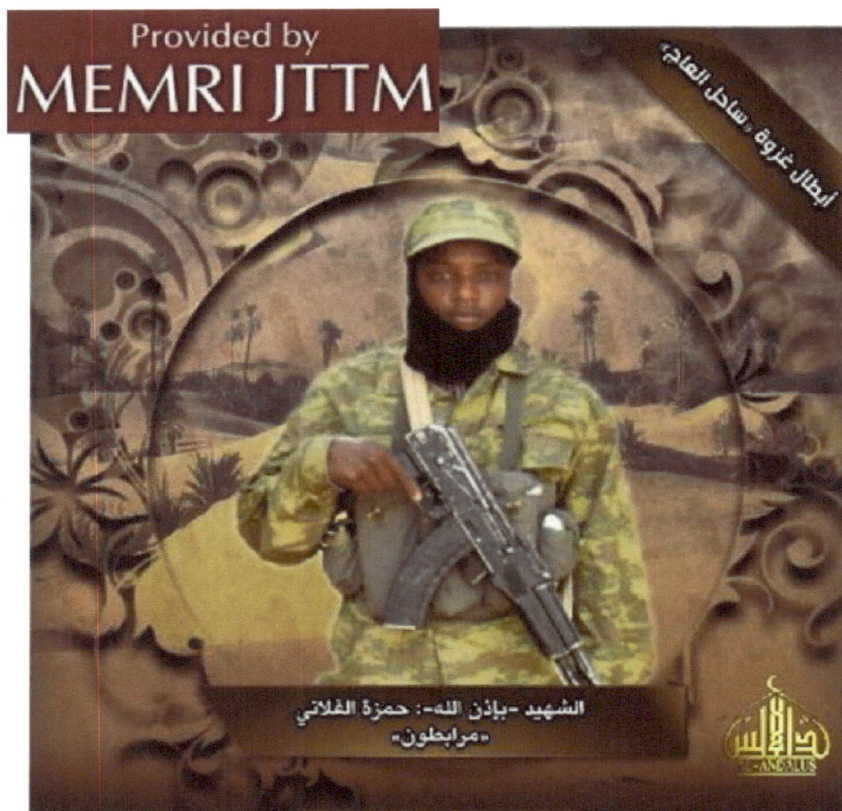

Hamza Al-Fulani

Al-Shabab Al-Mujahideen

In Audio Recording, Al-Shabab Leader Urges Somalis To Join Jihad, Calls Upon Muslims In Kenya, Ethiopia To Target "Unbelievers" In Any Way Possible, July 13, 2016

On July 12, 2016, Al-Shabab Al-Mujahideen, Al-Qaeda's Somalia branch, released a 44-minute audio recording of its leader Abu Ubaida, aka Ahmad Omar, titled "Shari'a or Martyrdom." The recording, released by on the Global Islamic Media Front (GIMF) Telegram channel, features Abu Ubaida praising Al-Shabab's jihad against their enemies, and calls upon them to intensify their attacks against the Somali government and the "invading Crusaders," a reference to the Ethiopian, Kenyan, and other Western-backed African troops, in Somalia. He also urges Somali youths to join the group, while calling upon Muslims in neighboring countries to send their children to the group's training camp, immigrate to Al-Shabab-controlled territory, or participate in lone wolf operations in their respective countries. He notes that Muslims continue to face a "fierce onslaught" from the "Crusaders, the Jews, the atheists and the apostates," who have imposed manmade laws and "constitutions of unbelief" upon Muslims and plundered their resources. However, Abu Ubaida says that he sees "signs of hope," depicted in a "worldwide jihad movement," which he says is being led by Muslim youths looking to reestablish the shari'a.

Abu Ubaida repeatedly extols jihad and calls on Muslims in the region to join it. He urges Muslims in Somalia to "stand up for jihad and fight against the Ethiopian and Kenyan Christians." Jihad, he says, "is the only source of your life and dignity." He sends a similar message to Somalis of all walks of life, urging them to "rise up and repel the conniving schemes" that are being devised against them and against the Muslim Ummah. "It is obligatory on you to wake up and realize the magnitude of this war," he says.

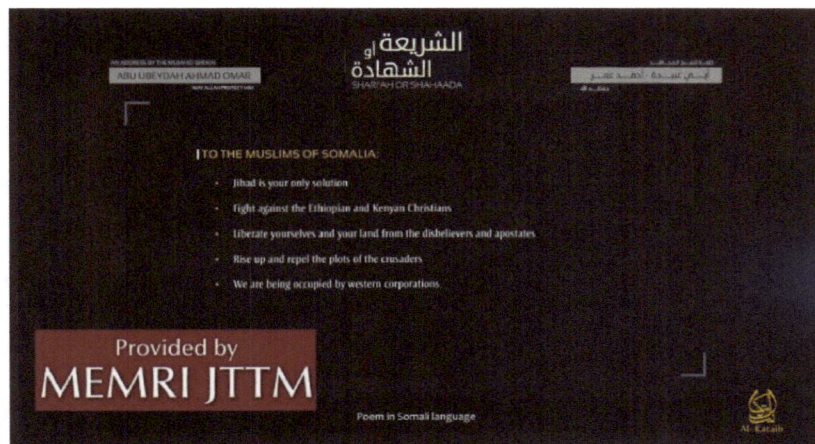

Abu Ubaida also addresses Somali youths in particular, urging them to join the jihad: "Know that whilst you are busy migrating from your lands, there are Ethiopian and Kenyan Crusaders of your age invading your lands and occupying your houses." Abu Ubaida expresses solidarity with Muslims in East and Central Africa, in particular in Ethiopia and Kenya, for the persecution they face by their respective governments and security services. He also promises them that Al-Shabab "will not spare any effort" in aiding them, avenging their deaths, and holding their enemies accountable: "We will not forget the blood of the Muslim scholars, which was spilt on the streets of Mombasa." Abu Ubaida also threatens anyone working with the Muslims' enemies there, or anyone involved in the killing of Muslims, informing them that they are "a target for us, and will soon pay a heavy price." Further, Abu Ubaida urges Muslims in those countries to send their children to Al-Shabab's training camps, and to immigrate to its territories. He also instigates them to carry out lone wolf attacks on the "disbelievers" in their countries by any means possible: "Take up arms and learn the different tactics of war... I encourage their youth to kill the disbelievers next to them and not reserve any energy in fighting them wherever they may be...Whoever can kill them using even a knife or an axe should do so. Whoever can use a vehicle to run them over should do so."[231]

Jabhat Al-Nusra/Jaysh Fath Al-Sham: U.S.-Designated Terrorist Cleric Mustafa Muhammad, Media Groups, Foreign Fighters

Australian Jabhat Al-Nusra Cleric – And U.S.-Designated Global Terrorist – Mustafa Muhammad Active On Telegram, July 18, 2016

Australian national Mustafa Muhammad, aka Abu Sulayman Al-Muhajir, who since May 19, 2016 has been a U.S. Treasury Department specially designated global terrorist, is a leading Jabhat Al-Nusra (JN) cleric and a member of the organization's shura council. Long active on social media, on June 5, Al-Muhajir created a Telegram channel which, as of this date, had 706 members. On Telegram, Al-Muhajir presents himself as accessible; interested parties can submit questions for him to answer on his channel.

Following a hiatus in his social media activity, Al-Muhajir returned to Twitter on June 6; he is also active on Facebook, where he created an account on July 14. His social media posts are devoted to attacking ISIS, discussing political or military events, and praising JN's efforts. Abu Sulayman maintains Twitter accounts in Urdu, English, German, and Turkish.

On July 6, the cleric lauded fighters' efforts: "To those who have lost a limb so others can keep theirs: #EidMubarak #HeroesOfJihad #NobleJihad." Previously, on June 6, Abu Sulayman conducted a poll on his Twitter page, on the question: "Over the past 5 years, the Syrian struggle has brought about more harm than good." Twenty-two percent of the 633 respondents agreed with the statement, while 78% marked the other option, "No. Jihad must continue."[232]

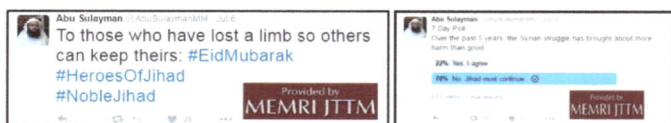

Abu Sulayman Praises The Syrians' Courage And Thanks Them For Their Hospitality, June 19, 2016

On June 17, 2016, the Al-Basira media group, affiliated with Jabhat Al-Nusra (JN), released a 12-minute video on the occasion of Ramadan titled "Giving Comes Naturally to the Loyalists." The video, which was posted on JN's official Telegram account, is an address in Arabic by Abu Sulayman Al-Muhajir, This is his first appearance in a JN video in more than a year.

In the video, Abu Sulayman praises the Syrians for courageously fighting both the Assad regime and the other enemies despite the severe blows they are dealt. He says that he has never seen anyone braver than the Syrians, who are willing to sacrifice their lives for Allah even though many of them have already lost their families, their property, and their homes.

Abu Sulayman also thanks the Syrians, and especially the jihad fighters among them, for welcoming the foreign fighters that have come to assist them: "I say in the name of all foreign fighters – those from Yemen, the Arabian Peninsula, the Islamic Maghreb, Egypt, Sudan, America, Turkistan, Uzbekistan, Chechnya, the Maldives, Britain, France, Venezuela, Germany, Belgium and elsewhere – that we are essentially living in your midst. I say in their name that nothing saddens me except the fact that we have nothing to give you [in return for your hospitality] – nothing but our souls that we will sacrifice for Allah in your blessed land. If we have to sacrifice our lives, one by one, so that you live in safety and no harm comes to you, we will not hesitate to do so. You made us feel that we are part of you and you are a part of us.."[233]

Pro-Jabhat Fath-Al Sham Media Group Showcases Kurdish, Tajik, Syrian Fighters On Telegram, October 17, 2016

The pro-Jabhat Fath-Al Sham (JFS) media group Fursan Al-Sham Media, which was launched in September 2016, disseminates its content on Telegram as well as on other social media. The group mainly eulogizes fighters, giving details on their backgrounds and on the battles in which they fought and also shares stories of current fighters. Fursan Al-Sham directs its Telegram followers to contact it via its Facebook page if they have any questions about Syria.[234]

On September 18, the group posted on Telegram the story of a current JFS fighter from Tajikistan: "If our enemy brings fighters from around the globe to fight Ahlul Sunnah [Sunnis] in Sham [Syria], we also have Muhajireen to migrate from the east to the west to Defend this Deen [religion]. Meet Abu Umar, a 36-year-old who made hijrah from Tajikistan 6 months ago leaving behind his son and wife. He is currently serving in JFS special forces and actively takes part in battles, he previously took part in the battle to free Aleppo and soon will be deployed to Lattakia. We ask Allah to preserve him and keep him steadfast."

Two individuals mentioned are quite young. On September 19, the story of "Abu Sufyan" was posted on Fursan Al Sham's Telegram account: "Ummah has a bright future ahead with the likes of this young man. This young Mujahid you see in the picture goes by the Kunya of Abu Sufiyan and he is from regime held Baniyas located in the Tartus governorate [Syria]. Once the blessed jihad reached the land of Al Sham Abu Sufiyan migrated from the Assad stronghold of Baniyas to the liberated areas of the Mujahideen along with his family to Idlib. Abu Sufiyan had no plans of sitting at home and playing with other children in his neighborhood considering he was of the age of 13, his enthusiasm to embark on jihad was incredible as he once tried to join Jabah Fatah Al sham by concealing his age to be 16.

"Not only did he want to join JFS but he had plans to enlist for the Inghamashi squad (front line stormers), once he reached one of the bases of JFS planning to go to one of the training camps he was prevented after some of the brothers found out his age to be 13, obviously this hurt Abu Sufiyan and he had no choice but to return back to his family. On the bright side Abu Sufiyan was told to return back after three months where he had the opportunity to enroll for an Islamic school to learn about the Deen and memorise the Quran, We ask Allah to reward him for his intentions and grow him up to be a fearsome Mujahid who will bring back glory to Islam. "

Left: Abu Sufyan. Right: Abu Bara

On September 22, the story of "Abu Bara" (above) was shared on Fursan Al Sham's Telegram account: "A family which continues to sacrifice for their Deen and Akhirah [afterlife]... The young 16 year old boy in the picture, who goes by the kunya of Abu Bara, has recently joined the caravan of the Mujahideen. Even though, two of his older brothers were martyred in separate battles, this did not demotivate him. At face-value, you would think that Abu Bara escaped from his family (as there is him, and only his younger brother to tend for their needs) for him to join the ranks, However, quite the opposite. As narrated by Abu Bara, in fact, the father himself encouraged Abu Bara to go forth for Jihad (and even threatened him with expulsion from the house if he refused to do so). So Abu Bara then expressed to me his delight of the prospect of him taking part in future battles, and he now also looks forward to participating in one of the training camps which has yet begun. We ask Allah to reward his family immensely for what they had to sacrifice and we ask Allah to keep Abu Bara steadfast on the path of Jihad. Ameen."

Hizbullah: Al-Manar TV Launches Telegram Channel

Designated Terrorist Organization Hizbullah's Al-Manar TV Joins Telegram, August 26, 2016

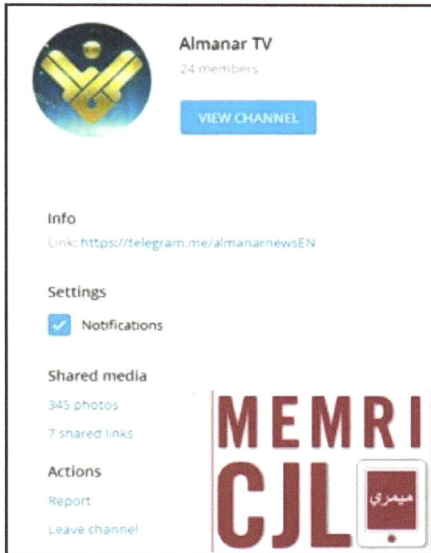

On August 26, 2016, Hizbullah's Al-Manar TV created its own Telegram channel; as of October 10, 2016, it had 25 members. The channel posts are mostly on developments by Hizbullah and Shi'ite groups and leaders in Iraq, Yemen, and especially Iran.

According to previous MEMRI research, Hizbullah's public relations efforts are aggressive and sophisticated; it is very active on social media with current accounts on Twitter, YouTube, and previously Facebook (over 10 accounts have been created and taken down over the past few years), and made attempts to create apps for Apple and Android devices.

Since Al-Manar TV began broadcasting via satellite in 2000, it has been at the center of controversy throughout the West. The channel has been banned in various capacities in the U.S., Canada, France, Australia, Spain, Netherlands, and elsewhere. While these countries have been successful in keeping Al-Manar TV off the airwaves, Hizbullah has managed to circumvent the ban with its multiple-language website, www3.almanar.com.lb, which live-streams its programming.

The following are some Telegram posts from Al-Manar's account.[235]

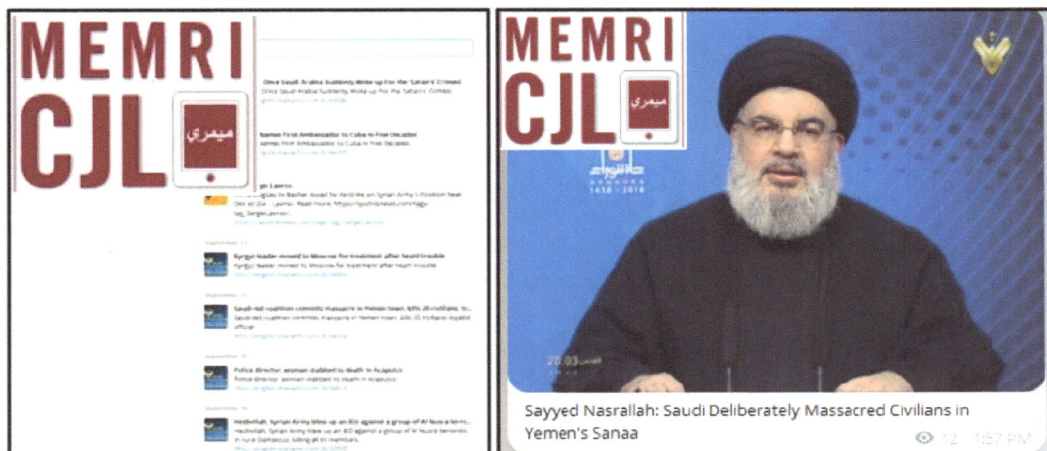

Sayyed Nasrallah: Saudi Deliberately Massacred Civilians in Yemen's Sanaa

Ayatollah Khatami: Saudi Criminals Should Go on Trial in Islamic Court

Hezbollah: US Allegations It's Fighting ISIL 'Propagandist'

Iraq's Sayyed Hakim: ISIL Will Be Crushed Soon

Rouhani Lands in Cuba for Talks with Castros

Commander: IRGC Ready to Counter All Threats

Leader Highlights Iran's Outright Distrust of US

US's Two Faces

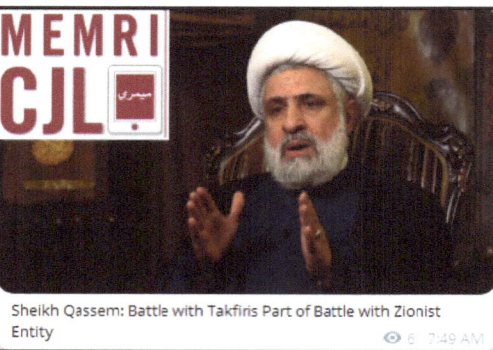
Sheikh Qassem: Battle with Takfiris Part of Battle with Zionist Entity

Taliban: News Updates, English-Language Channels, Official Statements

Taliban Telegram Channels Publish News Updates, January 19, 2016

Two English-language Taliban channels, Alemarah News and #Breaking Islamic Emirate, were created on Telegram in late 2015. The first went live on October 26, and the second was created on November 2, 2015. The two channels cover similar news stories, focusing on Taliban battles in Afghanistan and on American activity in the region. They also disseminate photos, audio broadcasts by leaders, and videos of battles. Both the accounts are loyal to the Islamic Emirate of Afghanistan, the Taliban shadow government headed by Taliban leader Mullah Akhtar Mansoor. The first post on the Alemarah News channel, on November 1, 2015, read: "35 soldiers killed as fighting erupted in locality of DashtiArchi district of Kunduz province." The post was accompanied by a photo from the scene of the fighting.[236]

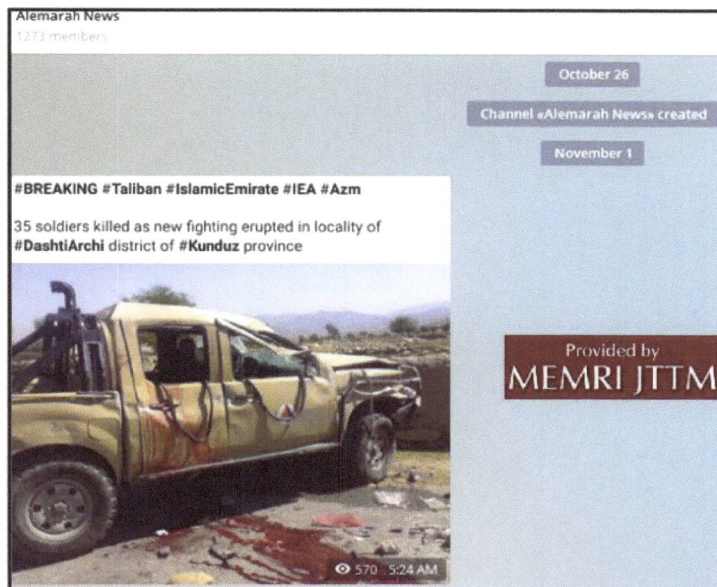

English-Language Taliban Telegram Channel Offers WhatsApp Updates Option, January 21, 2016

On December 10, 2015, the English-Language Taliban channel titled Al Emarah News offered its readers the option to receive group updates via the messaging service WhatsApp. An image posted to the channel included the phone number and username with which users could join the Al Emarah WhatsApp group. The message posted on Telegram read: "Glad tidings, All Muslim brothers can register their names with the following number to receive all Al Emarah releases in Pashto, Persian, English and Arabic languages. To register, send your name through Whatsapp to Al Emarah and receive breaking news and video releases of the Islamic Emirate."[237]

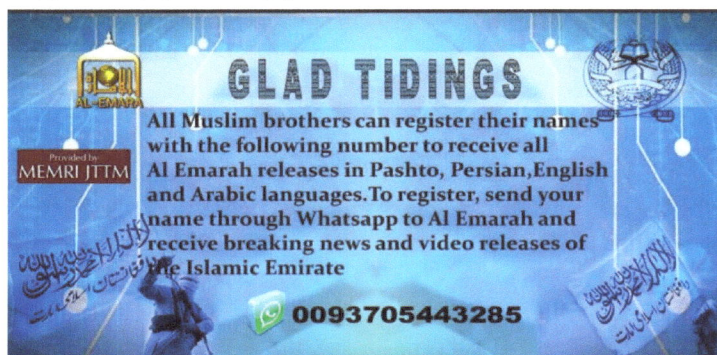

Taliban-Affiliated Telegram Account Solicits Donations For Eid Al-Adha, September 9, 2016

On September 6, 2016, the Taliban-affiliated Telegram account Alemarah-English Official posted a message asking for donations for Eid Al-Adha to help the needy.[238]

Alemarah-English Official
Forwarded from Alemarah-Enghlish Official (Asad Afghan)

AL-EMARA
دافغانستان اسلامي امارت

Admin: @asadafghan1

دالامارہ ستودیو رسمي کانال پہ تیلیگرام کې

Telegram
https://telegram.me/alemarastudio
0093708638285

Alemarah-English Official
Asad Afghan:

In the name of Allah, The most beneficent the most merciful

Allah almighty says وَيُطْعِمُونَ الطَّعَامَ عَلَى حُبِّهِ مِسْكِينًا وَيَتِيمًا وَأَسِيرًا
سورة الانسان، آية ٨.
Translation: And they give food in spite of love for Allah to the needy, the orphan, and the captive.
Muslim brothers! As you know, Islamic Emirate of Afghanistan is the guardian of thousands of widows and orphans and provide them sacrifice of Eid-ul-Azha. Therefore Islamic Emirate calls upon every fortune and sympathizer Muslims to share their best with the oppressed nation on this blessed occasion of Eid to change their grief with happiness.
In order to convey your sacrifices to the needy and war effected afghan nation contact the Islamic Emirate commission for financial affairs at the following email address, number. And if you have internet facility, the same number can be used for contact through Telegrams and WhatsApp. To perform your religious responsibly and to make the oppressed nation celebrate happy moments of Eid like your own family.
Thanks,
Islamic Emirate commission for financial affair
For contact:
009305790096
Financecm.iea@gmail.com

On Telegram, Taliban Lashes Out At Brussels Conference On Afghanistan, October 6, 2016

On October 4, 2016, the Taliban's official Telegram channel posted a rebuke of the October 4-5, 2016 conference held in Brussels and hosted by the European Union and the government of Afghanistan, which included over 70 governments and 30 international organizations and agencies. The Taliban statement explains that ideally, it would like to see the participants in the conference call for the complete withdrawal of all foreign troops from Afghanistan, and insists that this would be the most beneficial course of action for Afghans: "In reality the occupying countries – due to their colonial peculiarity – always seek to prolong their occupation and attain their colonial goals through such conferences. This is because they always give precedent to their interests over the wellbeing and welfare of the people."[239]

Taliban Issues Statement, Threats Of U.S. Invasion Of Afghanistan, October 11, 2016

On October 7, 2016, the Taliban's official Telegram channel posted a "Statement of Islamic Emirate Regarding 15th Anniversary of American Invasion In Statements." According to the statement, because of the U.S. presence in Afghanistan, the country is now one of the poorest nations in the world, and despite its abundant natural resources, it has advanced very little. The statement also blames the country's rampant government corruption and instability on the American presence, and concludes with an ultimatum to the Americans in Afghanistan to leave, or else "the believing Afghan nation will continue their legitimate struggle under the leadership of the Islamic Emirate, until the invaders are expelled from the country like the previous ones, [w]hen you will have lost tens of thousands more troops and wasted hundreds of billions of more dollars in exchange for a historical and humiliating defeat."[240]

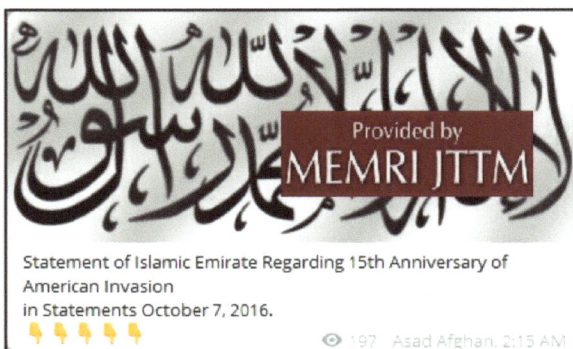

Statement of Islamic Emirate Regarding 15th Anniversary of American Invasion
in Statements October 7, 2016.

Turkestan Islamic Party (TIP): Urging Xinjiang Muslims To Join Jihad, Multiple Channels, News Of TIP Fighters In Aleppo Battles; Jihadi Music Videos

Turkestan Islamic Party (TIP) Launches Telegram Channel, January 13, 2016

On November 24, 2015, the Syrian branch of the Turkestan Islamic Party (TIP), an Al-Qaeda affiliate comprising mainly Uyghur fighters, launched a channel on the encrypted messaging app Telegram. Official photos and videos, as well as promotions for new productions, are circulated on the channel. The channel is named after TIP's media wing Sawt Al-Islam.[241]

Turkestan Islamic Party (TIP) In Syria Urges Xinjiang Muslims To Join The Jihad In New Video, July 12, 2016

On July 11, 2016, the Syrian contingent of the Al-Qaeda-affiliated Turkestan Islamic Party (TIP) released a new video urging Uyghur Muslims in China's Xinjiang region to join the ranks of the mujahideen in Syria. The five-minute video, titled "Come to Jihad," was released by the TIP media branch, Islam Awazi, on Telegram. The video is in Uyghur with subtitles in both Arabic and English. The video opens and closes with a *nasheed* [religious chant] in Uyghur, with background images of Uyghur fighters training and carrying out military operations in Syria. The video also shows many images of China's Xinjiang province, aka East Turkestan. The video calls for the East Turkestan separatists to give up political activism and take up arms to fight the jihad. The lyrics of the song are as follows: "Come to jihad O oblivious ones. Come forth, you who are in search of baseless excuses. Come to jihad, you who are unconcerned with an eternal paradise that is as vast as the heavens and the earth. Our creed and our belief is our weapon."[242]

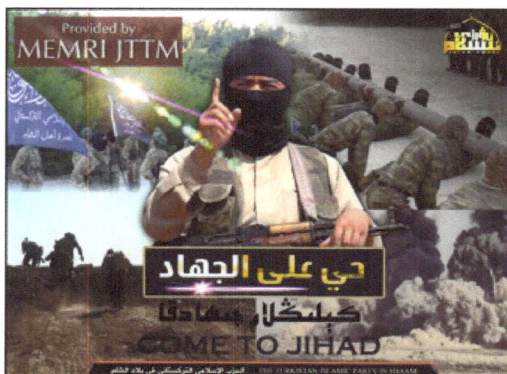

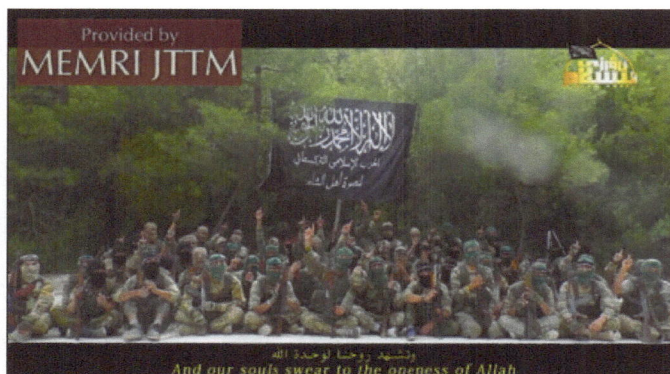

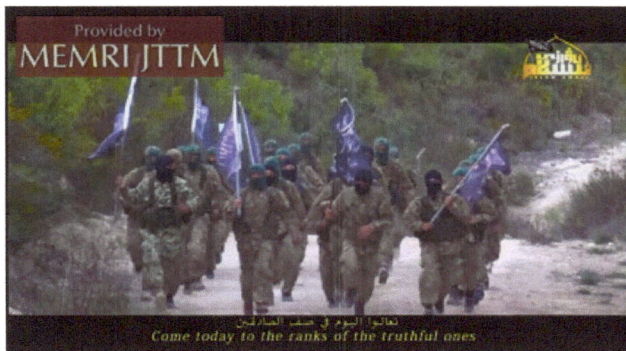

Come today to the ranks of the truthful ones

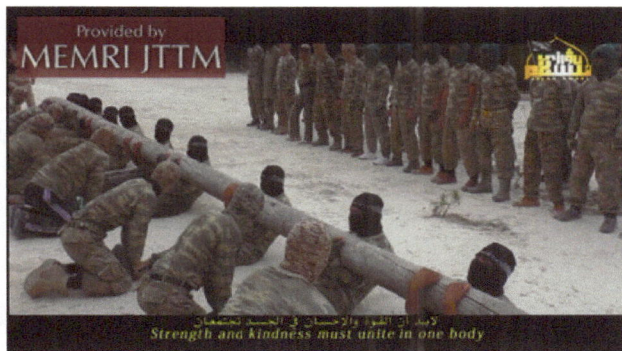

Strength and kindness must unite in one body

Turkestan Islamic Party (TIP) Launches New Channel On Telegram, August 10, 2016

On August 2, 2016, Sawt Al-Islam, the media arm of the Turkestan Islamic Party (TIP), launched a new channel on Telegram covering the activities of the group's branch in Syria. A week later, the channel had nearly 3,000 followers, and was being used by the Al-Qaeda affiliated group to post photos and videos documenting its ongoing fighting against the Assad regime and its allies in Syria.[243]

According To Its Telegram Account, TIP Fighters Heavily Involved In Aleppo Battles, August 10, 2016

Uyghur fighters of the Turkestan Islamic Party (TIP) contingent in Syria (which now calls itself "The Turkestan Islamic Party for Supporting the People of Al-Sham") have been heavily involved in the recent battles in the city of Aleppo and in the rebels' effort to break the siege of the pro-Assad forces. This is evident from several statements issued by TIP since July 31, which presented details of its operations in the area and photos of its members participating in the fighting. According to these statements, published on TIP's Telegram channel, TIP fighters took part in artillery, mortar and rocket attacks on Syrian army positions and in storming the pro-Assad forces. The photos showed weapons and ammunition seized by its fighters and the bodies of Syrian soldiers they killed.

Turkestan Islamic Party (TIP) Releases New Music Video In Uyghur: "Rise Up, Oh Turkestan," September 6, 2016

On September 2, 2016, the Turkestan Islamic Party (TIP) in Syria released a new music video of a nasheed (Islamic song) on its official media channel on Telegram "Voice of Islam." The nasheed calls on the Uyghur people to rise and take up arms against the oppression of the Chinese government. The 4:30-minute video, titled "Rise Up, Oh Turkestan," is in the Uyghur language with English and Arabic subtitles.[244]

ISIS Releases New Song In Uyghur, Titled "Religion Of Ibrahim," August 24, 2016

On August 22, 2016, ISIS's Al-Hayat media center, which is responsible for media production in non-Arabic languages, distributed on its Telegram channel a new Islamic chant in the Uyghur language, titled "Religion of Ibrahim." Music plays an important role in recruitment and spreading ISIS's message.[245]

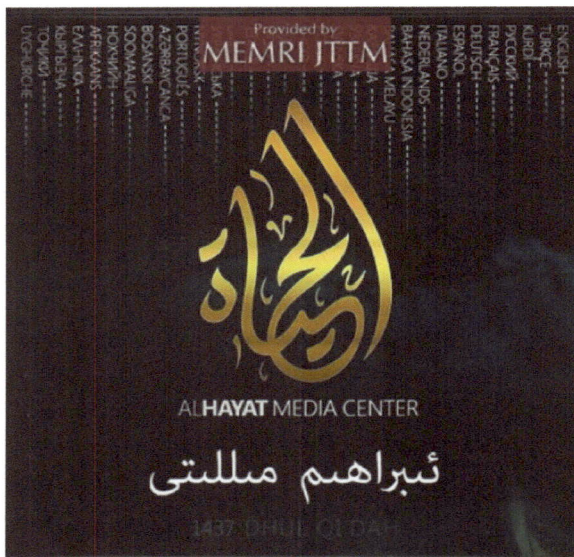

Hamas And Gaza: Al-Qassam Martyrs Brigade Promotes Jihad And Martyrdom, Manuals For Stabbing Attacks, Bomb-making; Fundraising For Weapons To Target Jews

On Its Telegram Account, Hamas's Al-Qassam Martyrs Brigades Stresses Jihad And Martyrdom, March 29, 2016

On its Telegram channel, which it opened in November 2015, the Izz Al-Din Al-Qassam Martyrs Brigades, the military wing Hamas, shares photos, details its exploits – battles, military parades, and terror attacks – and disseminates anti-Israel propaganda.[246]

Telegram Account Reposts Document Offering Stabbing Tactics, Bomb Making Instruction For Palestinians Planning On Attacking Jews, January 26, 2016

On January 26, 2015, a jihadi Telegram account posted a document with instructions and tips on stabbing tactics, as well as bomb and poison manufacturing, for those planning to kill Jews. The document was originally published in October 2015 by the Ibn Taymiyya Media Center (ITMC), which features news on jihadi groups in Gaza. The stated aim of the document, which is titled "Jihadi Ideas and Tips," is to maximize the results of individual attacks, as well as to incite and support the "lions of individual jihad operations" in Bait Al-Maqdis (Jerusalem).[247]

On Telegram, A Campaign To Fund Weapons, Fighters To Target Jews, November 13, 2015

A campaign was launched on Telegram as well as on Twitter to fund terrorist attacks against Jews in Israel. The "Al-Aqsa Nafeer Campaign," launched in October 2016, states that it aims to "prepare the mujahideen of Bait Al-Maqdis [Jerusalem, also used to designate Palestine] … [to carry out] jihadi and martyrdom operations targeting the Jews…" It also says that the attacks are aimed at "liberating the imprisoned Al-Aqsa Mosque from the filth of the aggressing Jews." This is not the first such campaign online; a campaign by the same name launched in July 2015 sought donations to support the families of Palestinian "martyrs" and prisoners, and collected food for the needy during Ramadan. However, the current campaign is geared towards collecting funds for weapons and other military gear needed by the mujahideen in Palestine in order to target the Jews; so far, according to a statement released by the campaign, it has been "successful."[248]

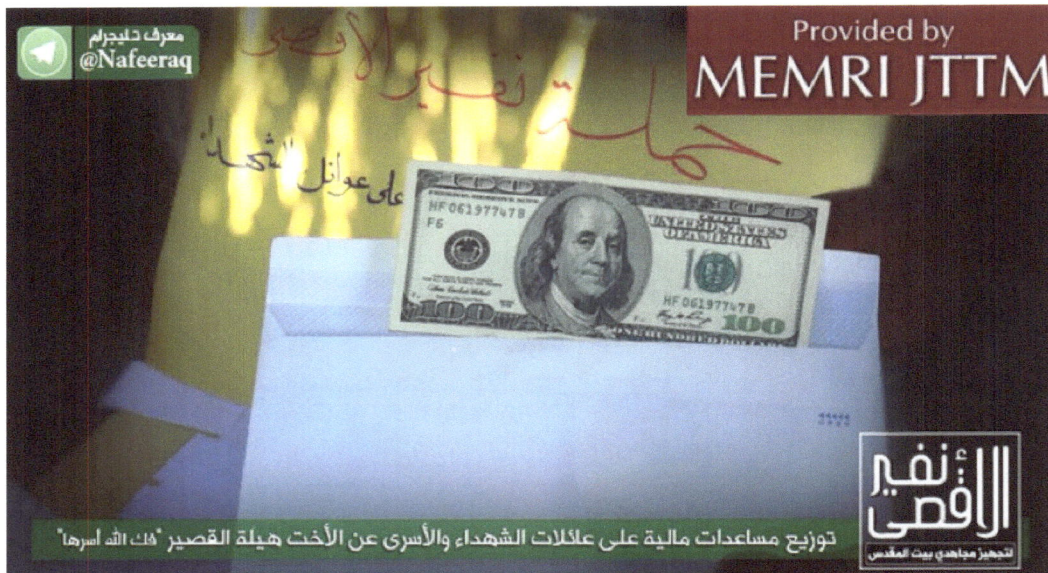

Funds collected for the families of Palestinian martyrs and prisoners. (Source: Justpaste.it/mnce, July 28, 2015)

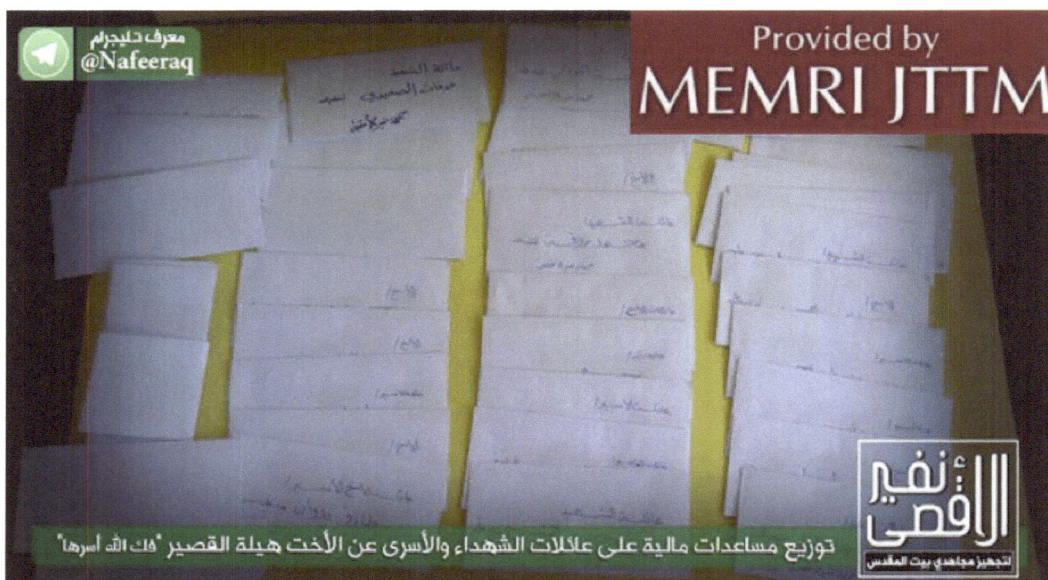

Envelopes with money addressed to prisoners' and martyrs' families. (Source: Justpaste.it/mnce, July 28, 2015)

Fundraising Campaign To Arm Jihadis In Gaza Solicits Donations Via Bitcoin, July 8, 2016

The "Jahezona" ("Equip Us") campaign, which has been running for almost a year, and aims to raise funds for the mujahideen in Gaza for weapons purchasing and development, has recently begun soliciting donations in the cryptocurrency Bitcoin, primarily via a Telegram campaign. The campaign uses the platform to post calls for donations, promotional posters, weapons pricelists, and incitement against Jews, and http://www.memrijttm.org/fundraising-campaign-to-arm-jihadis-in-gaza-solicits-donations-via-bitcoin.html - _edn2 was launched in July 2015 by the Ibn Taymiyya Center (ITC), a media body affiliated with Salafi-jihadi groups in Gaza.

Poster soliciting donations for the mujahideen in Gaza; poster from the campaign placing the "cost for equipping a mujahid" at $2,500

The first solicitation for Bitcoin donation was posted on June 29 on the campaign's Telegram account. stating: "In addition to the secure means through which we receive your donations, you can send donations via #bitcoin using the attached [bitcoin wallet address] code."[249]

A price list for various rockets showing their "effectiveness" in targeting Jews and Jewish settlements, along with the bitcoin address QR Code (red arrow)

Other Jihadi Activity On Telegram – Internet Archive, Nasheeds, Well-Known Jihadi Figures

Other Well-Known Jihadis On Telegram

The Telegram channel of Palestinian-Jordanian pro-Al-Qaeda jihadi Abu Qatada, who was deported by the UK to Jordan, was launched on August 27, 2017, and as of October 14, 2016 had 4,484 members. Egyptian Salafi-jihadi Dr. Hani Al-Siba'i, who heads the Al-Maqrizi Center for Historical Research in London, launched a Telegram channel November 7, 2015, and as of October 14, 2016 it had 2,320 members. Senior Jabhat Fath Al-Sham religious figure Abdallah Muhaisny's Telegram channel was launched September 29, 2015, and as of October 14, 2016 had 35,023 members.

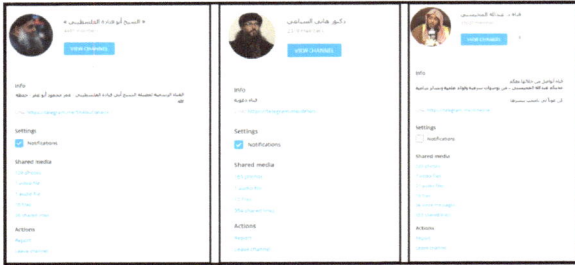

Left to right: Telegram channels of Abu Qatada, Hani Siba'i, and Abdallah Muhaisny.

Sheikh Abu Muhammad Al-Maqdisi, the prominent Jordan-based Salafi spiritual leader, who was once advisor to Al-Qaeda in Iraq founder Abu Musab Al-Zarqawi, launched his Telegram channel on September 26, 2016. As of December 14, 2016, it had 1,017 members, and had posted 15 videos, one video file, two other files, and nine share links. When his website Minbar Al-Tawhid Wal-Jihad, which had served as a primary means of disseminating Salafi-jihadi doctrine and jihad-related fatwas, was shut down two years ago, he began using Twitter, and has now expanded to Telegram. At its launch, the account posted: "We shall be providing translations of important tweets/quotes/messages of Sheikh Abu Muhammad al Maqdisi in English in sha Allah. Please share and spread. We ask Allah to accept our efforts." The account answers questions, for example, "is jihad more important, or Dawa [outreach]?..." The answer to this question was, "None of them is sufficient without the other." It also posts quotes by Islamic scholars, such as by the 14th-century Islamic jurist Muhammad ibn Abu Bakr (Ibn Al-Qayyin), who was quoted as saying: "Giving victory to this religion is a necessary obligation... compulsory on every single Muslim."

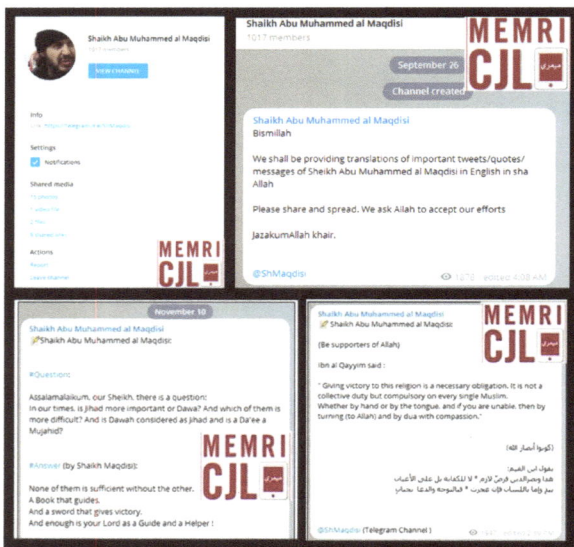

Sheikh Abu Muhammad Al-Maqdisi's Telegram account with sample posts.

Jihadi Telegram Channels Using Internet Archive To Spread Content From ISIS, Al-Qaeda, And Other Groups, November 3, 2016

In 2011, MEMRI began monitoring jihadi groups utilizing the Internet Archive (Archive.org) to spread their propaganda, and the site has proven a jihadi favorite, as content is rarely removed from the site.[250] According to its website, the Internet Archive is "a 501(c)(3) non-profit that was founded to build an Internet library" by cataloguing files and web pages that would otherwise no longer be accessible to Internet users.[251] Online jihad supporters have been utilizing the Internet Archive as a file storage database and a venue for distributing their propaganda materials, as well as applications and file-sharing links, which are then shared by Telegram channels promoting the messages of jihadi groups including ISIS and Al-Qaeda.[252] Below are examples of files in multiple media formats of media posted on the Internet Archive and linked to by jihadi Telegram channels. File types include videos, audio fles, PDFs, and file-sharing torrents.

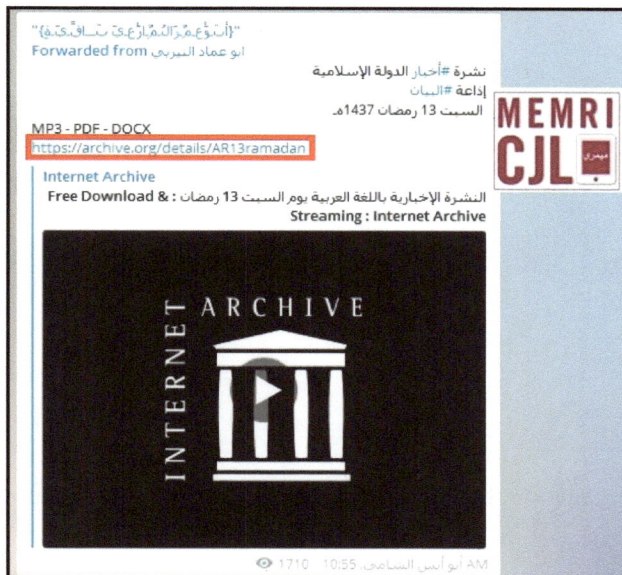

Video Files And Pictorial Reports

This channel shared a video playlist by Al-Hussam Al-Zarqawi Foundation, which includes various pro-ISIS and Al-Qaeda montages:

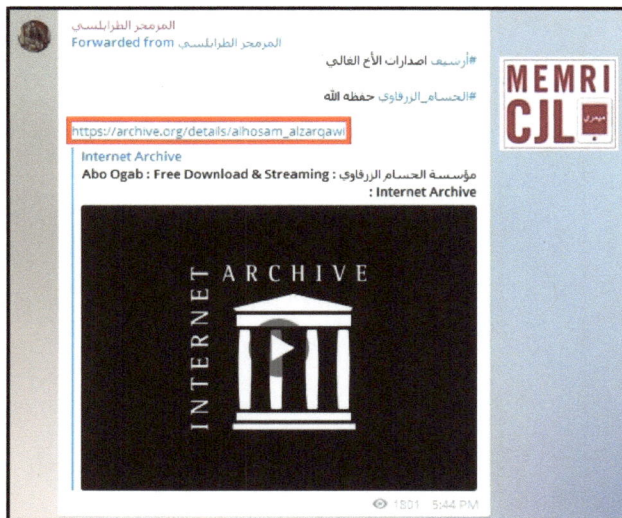

The Telegram channel Islamic State Productions shared a promotional video for an ISIS app that teaches children supplications:

This video, commemorating the shooting attack at the Pulse nightclub in Orlando, Florida, by Omar Mateen, was shared by the Telegram channel Updates from the Islamic State:

A link to an Islamic State Al-Bayan radio broadcast; these are posted daily in multiple languages:

Pictorial report showing the execution of Iraqi soldiers by ISIS:

"They Might Cease" video, shared by the channel:

Telegram "Nasheed Gallery" Channel Shares Songs Featuring Voices Of Jihadis, "Martyrs," July 11, 2016

The Telegram channel Nasheed Gallery, created June 30, 2016, allows users to listen to and download jihadi nasheeds (Islamic songs). According to the channel's description, the nasheeds are sung by "mujahideen and shuhadaa [martyrs]" and "those who are truthful in conveying the message of jihad." The channel encourages users to listen to the nasheeds while driving their car or "when you need motivation." The channel currently features over 50 nasheeds that can be streamed on Telegram or downloaded to a user's device. The nasheeds convey an array of messages pertaining to jihad, martyrdom, and fighting Islam's enemies. ISIS's nasheeds are a main factor in reinforcing the group's narrative and attracting new recruits.[253]

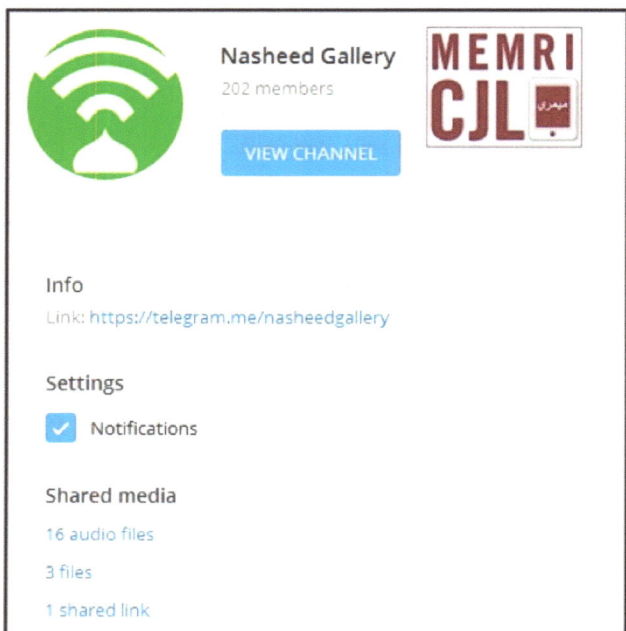

Egyptian Militant Group Jama'at Al-Murabitoon Uses Telegram, October 5, 2015

The Egyptian militant group Jama'at Al-Murabitoon has launched an account Telegram, according to an announcement that was posted on the Al-Fida' forum. People can use Telegram to contact the Islamic Shari'a Soldiers Brigade, Al-Murabitoon's supposed military wing, according to the announcement.[254]

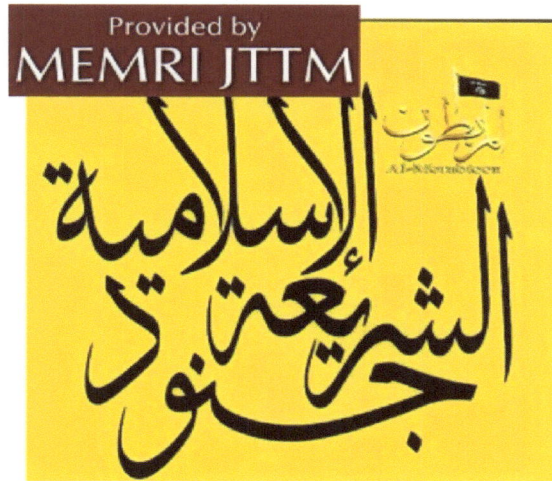

Steven Stalinsky is Executive Director of MEMRI; R. Sosnow is Head Editor at MEMRI; M. Al-Hadj is Director of the MEMRI Reform Project; M. Khayat, A. Agron, R. Green, S. Benjamin, and M. Shemesh are Research Fellows at MEMRI.

Endnotes

1   See also MEMRI CJL report Encryption Technology Embraced By ISIS, Al-Qaeda, Other Jihadis Reaches New Level With Increased Dependence On Apps, Software – Kik, Surespot, Telegram, Wickr, Detekt, TOR: Part IV – February-June 2015, June 16, 2015; Inquiry & Analysis Series Report No. 1143, Al-Qaeda's Embrace Of Encryption Technology Part III – July 2014-January 2015: Islamic State (ISIS) And Other Jihadis Continue To Develop Their Cyber And Encryption Capabilities; Post-Snowden Fears Lead Them To Test New, More Secure Technologies And Social Media, February 4, 2015; Inquiry & Analysis Series Report No. 1086 Al-Qaeda's Embrace Of Encryption Technology – Part II: 2011-2014, And The Impact Of Edward Snowden, April 25, 2014; and Inquiry & Analysis Series Report No. 704 Al-Qaeda's Embrace of Encryption Technology – Part I: 2007-2011, July 12, 2011.

2   See MEMRI CJL report Encryption Technology Embraced By ISIS, Al-Qaeda, Other Jihadis Reaches New Level With Increased Dependence On Apps, Software – Kik, Surespot, Telegram, Wickr, Detekt, TOR: Part IV – February-June 2015, June 16, 2015.

3   Washington Post, October 29, 2015.

4   Washington Post, October 16, 2014.

5   See MEMRI CJL report Encryption Technology Embraced By ISIS, Al-Qaeda, Other Jihadis Reaches New Level With Increased Dependence On Apps, Software – Kik, Surespot, Telegram, Wickr, Detekt, TOR: Part IV – February-June 2015, June 16, 2015.

6   See MEMRI JTTM report Pro-ISIS Writer Urges Online Jihadis To Defy Twitter Suspension Policy, Warns Against Switching To Telegram, October 5, 2016.

7   See MEMRI JTTM report In Telegram Secret Chat ISIS Supporters Plan Attacks In U.S., U.K., Aim To Target U.S. Embassy In U.K., November 14, 2016.

8   MEMRI Daily Brief No. 72, "Supporters Of The Islamic State" – Anatomy Of A Private Jihadi Group On The Encrypted 'Telegram' App, Offering Secret Chats And Private Encryption Keys, January 8, 2016.

9   See MEMRI CJL report Jihadis On Telegram Warn Against New WhatsApp Encryption Feature, April 14, 2016.

10   Washington Times, August 2, 2016.

11   Cnn.com, August 1, 2016; Bbc.com, August 10, 2016; New York Times, September 5, 2016.

12   The Washington Post, July 19, 2016.

13   See MEMRI JTTM report Top ISIS Propagandist 'Turjiman Al-Asawirti' Launches Telegram Channel, February 10, 2016.

14   Scmp.com, October 29, 2016.

15   Desktop.telegram.org, accessed August 23, 2016.

16   Mirror.co.uk, May 17, 2016.

17   Businessinsider.com, November 21, 2015.

18   Money.cnn.com, December 17, 2015.

19   See MEMRI Inquiry & Analysis Series report No. 1118 Wives Of Foreign Jihad Fighters In Syria – Active Members Of The Jihad Community, September 18, 2014; Inquiry & Analysis Series report No. 1100 Russian Jihadis Use Social Media To Raise Funds For Jihad In Syria, June 23, 2014; MEMRI Daily Brief - No. 100, History Will Judge Tech Philanthropists In Fight Vs. Cyber Jihad, August 22, 2016; MEMRI JTTM report Message On Russian Social Network: Muslim Shishani's Jihad Group Has Been Attacked With Chemical Weapons, March 13, 2014.

20   Money.cnn.com, December 17, 2015:

21   Telegram.org/blog/cryptocontest-ends

22   Businessinsider.com, November 21, 2015.

23   Telegram.org/faq, accessed August 24, 2016.

24   Armed-services.senate.gov/hearings/16-09-13-encryption-and-cyber-matters.

25   Telegram.org/faq, accessed August 23, 2016.

26   Businessinsider.com, November 21, 2015.

27   Telegram.org/faq, accessed August 24, 2016.

28   Telegram.org/faq, accessed August 23, 2016.

29   Businessinsider.com, November 21, 2015.

30   C-span.org, December 7, 2015.

31   New York Times, September 5, 2016.

32   Cnn.com, August 1, 2016.

33   Cbsnews.com, March 13, 2016.

34   New York Times, September 5, 2016.

35   Volkskrant.nl, September 17, 2016.

36   Youtube.com/watch?v=kVZN9QbtFgs, September 23, 2015.

37   Techchrunch.com, September 21, 2015.

38   Techcrunch.com, September 21, 2015.

39   See MEMRI JTTM report Jihadis Shift To Using Secure Communication App Telegram's Channel Service, October 29, 2015

40   See MEMRI CJL report Arab News Site Claims 150 Terrorists Arrested Through Telegram, November 6, 2015. It should be noted that the MEMRI JTTM team identifies numerous issues with this story – first, it refers to the terrorist activists as belonging to Ansar Bait Al-Maqdis, a name that the group no longer uses since it became ISIS Sinai after swearing fealty to Abu Bakr Al-Baghdadi. Second, the story says that the group itself announced the arrest, although MEMRI can find no record of this. Third, the story refers to the group's leader as Abu Sayyaf Al-Masry, even though this name is not known in connection with the group.

41   Cbsnews.com, March 13, 2016.

42   Nytimes.com, November 18, 2015; MEMRI CJL report Telegram Blocks Dozens Of ISIS-Related Channels For The First Time, November 18, 2015.

43   See MEMRI CJL report Telegram Blocks Dozens Of ISIS-Related Channels For The First Time43   Cnn.com, December 17, 2015.

44   Cbsnews.com, March 13, 2016.

45   Cbsnews.com, March 13, 2016.

46   Timesofindia.com, July 25, 2016.

47   Wsws.org, February 1, 2016.

48   Wsj.com, January 16, 2016.

49   Freepressjournal.in, July 21, 2016.

50   Thehindu.com, July 1, 2016.

51   Wsj.com, July 23, 2016.

52   The Washington Post, July 19, 2016.

53   Cnn.com, August 1, 2016.

54   BBC.com, August 10, 2016; New York Times, September 5, 2016.

55   Filipinotimes.ae, August 11, 2016.

56   Themoscowtimes.com, November 16, 2015.

57   Nytimes.com, November 18, 2015.

58   Venturebeat.com, November 19, 2015.

59   Dailydot.com, November 19, 2015.

60   Independent.co.uk, November 20, 2015.

61   Telegramgeeks.com, March 2, 2016.

62   Yahoo.com, February 23, 2016.

63   Cnn.com, February 23, 2016.

64   Fortune.com, February 23, 2016.

65   Theverge.com, February 23, 2016.

66   Ft.com, July 3, 2015.

67   Cnn.com, February 23, 2016.

68   New York Times, September 5, 2016.

69   Scmp.com, October 29, 2016.

70   Telegram.org/faq#q-there-39s-illegal-content-on-telegram-how-do-i-take-it-down

71   At that link, under the heading "Q. There's illegal content on Telegram. How do I take it down?" it stresses that "All Telegram chats and group chats are private amongst their participants. We do not process any requests related to them" and then goes on to discuss "legitimate requests to take down illegal public content" [emphasis in original] without stating how to submit them. Nor under a subsequent heading, "Q. Wait! Do you process take-down requests from third parties?" does it provide clear instructions for submitting requests, although it says: "Whenever we receive a complaint at abuse@ telegram.org or dmca@ telegram.org regarding the legality of public content, we perform the necessary legal checks and take it down when deemed appropriate." Then, as mentioned previously in this report, it stresses that this "does not [emphasis in original] apply to local restrictions on freedom of speech" such as criticism of a government where it is illegal "because this "goes against our founders' principles," it adds that "[w]hile we do block terrorist (e.g. ISIS-related) bots and channels, we will not block anybody who peacefully expresses alternative opinions."

72   Washingtontimes.com, August 2, 2016.

73   See MEMRI JTTM report Jihadis Announce New ISIS App For Android, August 3, 2015.

74   See MEMRI Inquiry & Analysis No. 1168, Encryption Technology Embraced By ISIS, Al-Qaeda, Other Jihadis Reaches New Level With Increased Dependence On Apps, Software – Kik, Surespot, Telegram, Wickr, Detekt, TOR: Part IV – February-June 2015, June 15, 2015.

75   BBC.com, September 15, 2016; WSJ.com, October 20, 2016.

76   See MEMRI JTTM report Prior To Attack, Adel Kermiche, Killer Of French Priest, Wrote On Social Media: 'In A Very Short While... There Will Be Big Things [i.e. News] On This Page', July 26, 2016.

77   See MEMRI report The French-Language Media Entity Ansar At-Tawhid And Its Role In ISIS Terror Activity – Part I: Operational Aspects, Indications Of Influence On Larossi Abballa, July 15, 2016.

78   See MEMRI JTTM report Rachid Kassim, French ISIS Operative And Executioner Active On Telegram – Threatens And Incites Terror, August 22, 2016.

79   BBC.com, September 15, 2016.

80   See MEMRI JTTM report ISIS Video Celebrates Nice Attack, Threatens Further Attacks Across France, Including In Paris, Marseille, And Nice, July 20, 2016.

81   See MEMRI JTTM video Before Attack in Nice, ISIS Fighter Calls to Carry Out Lone-Wolf Attacks in France: Go Get a Truck, July 19, 2016.

82   See MEMRI JTTM report ISIS Operatives Distribute Kill List, Target List, Operational Tips For 'Lone Lions' In France And The West, August 15, 2016.

83   See MEMRI JTTM report Rachid Kassim, French ISIS Operative And Executioner Active On Telegram – Threatens And Incites Terror, August 22, 2016.

84   See MEMRI JTTM report French ISIS Activist Rachid Kassim Calls On Young Algerians To Attack Tourists And Destroy Tourist Attractions, September 22, 2016

85   See JTTM MEMRI report ISIS Operatives Distribute Kill List, Target List, Operational Tips For 'Lone Lions' In France And The West, August 15, 2016.

86   See MEMRI JTTM report Rachid Kassim, French ISIS Operative And Executioner Active On Telegram – Threatens And Incites Terror, August 22, 2016.

87   See MEMRI JTTM report Chinese And Norwegian ISIS Prisoners For Sale In Issue 11 Of ISIS Magazine Dabiq; Available For Purchase Through Encrypted App Telegram, September 10, 2015.

88   See MEMRI JTTM report ISIS Video Sets Out Structure Of Caliphate State, July 6, 2016.

89   MEMRI JTTM report UPDATED – ISIS Announces The Death Of Spokesman Abu Muhammad Al-Adnani In Syria, Promises Retaliation, August 30, 2016.

90   See MEMRI JTTM report In ISIS Video, French Children Threaten: "Today In Syria, Tomorrow In Paris," May 17, 2016.

91   See MEMRI JTTM report ISIS Claims Responsibility For Attack On Train In Germany, July 19, 2016.

92   See MEMRI JTTM report ISIS Claims Responsibility For Suicide Attack In Ansbach, Germany, July 25, 2016.

93   See MEMRI JTTM report ISIS Releases Video Of Second Normandy Church Attacker, Who Calls On Muslims To Attack France, Coalition Countries, July 28, 2016.

94   See MEMRI JTTM report ISIS Releases Video Of Normandy Church Attackers Pledging Allegiance To Al-Baghdadi, July 27, 2016.

95   See MEMRI JTTM report ISIS Claims Responsibility For Copenhagen Attacks, Says Attacker Responded To Group's Call To Target Coalition Countries, September 2, 2016.

96   See MEMRI JTTM report ISIS Claims Responsibility For Attack On Russian Policemen, Releases Video Of Attackers, August 19, 2016.

97   See MEMRI JTTM report ISIS Appeals For Support In Philippines, Indonesia, And Malaysia, Calls On Its Supporters There To Attack "Unbelievers' And 'Apostates," June 22, 2016.

98   See MEMRI JTTM report ISIS Infographic Shows Its Areas Of Operation, Boasting Presence In France, Turkey, Many More Countries, June 29, 2016.

99   See MEMRI JTTM report ISIS Video Shows Fighters Deployed In Mosul At Night; Fighter Threatens America: "You Will Be Defeated Again In Iraq And Leave It Humiliated With Your Tail Between Your Legs," October 18, 2016.

100   See MEMRI JTTM report Top ISIS Propagandist 'Turjiman Al-Asawirti' Launches Telegram Channel, February 10, 2016.

101   See MEMRI JTTM report Extremist British Preacher Abu Haleema Launches Telegram Channel, Which Is Then Promoted By ISIS Activists, February 29, 2016

102   See MEMRI JTTM report Top ISIS Writer Asks Twitter, Telegram To Halt Their Suspension Of Jihadi Accounts; Challenges Them, As Well As Anti-ISIS Groups, To Counter-Argue, February 12, 2016.

103   See MEMRI JTTM report Jihadi Writer Warns ISIS Supporters To Not Limit Their Activities To Telegram, Urges Them To Use Facebook, Twitter, August 11, 2016.

104   See MEMRI JTTM report On His Telegram Channel, British ISIS Fighter Appeals For Donations To Keep Fighters Warm In Syria, December 14, 2015.

105   See MEMRI JTTM report "Islamic State In Libya" Telegram Channel Spreads 'Awareness Of The Expansion' Of The Caliphate In Libya, February 19, 2016.

106   See MEMRI JTTM report Pro-ISIS Media Group Releases Video Calling For Attacks In U.S., France, Belgium, Italy, Denmark, Spain, Russia And Iran, August 17, 2016.

107   See MEMRI JTTM report ISIS Activist Promotes His 'Online Da'wah Attack Operations' Telegram Channel On Facebook, February 26, 2016.

108   See MEMRI JTTM report On Telegram, American ISIS Member In Syria Promotes Belgian Recruiter To Her Followers, March 21, 2016.

109   See MEMRI JTTM report 'Shoot Obama' Game Shared In Pro-ISIS Telegram Group, April 1, 2016.

110   See MEMRI JTTM report American Who Claims To Be An ISIS Fighter Spreads Propaganda Via Telegram Group, Twitter, May 10, 2016.

111   See MEMRI JTTM report Female German ISIS Members Recruit On Facebook And Telegram, Share Life Experience Under ISIS Rule, May 24, 2016.

112   See MEMRI JTTM report Pro-ISIS Telegram Channel Threatens Christians In Tripoli, Lebanon, July 12, 2016.

113   See MEMRI JTTM report Pro-ISIS Media Company Celebrates Orlando Attack: "America – The Moment Of Your End Has Come," June 16, 2016.

114   See MEMRI JTTM report Swedish ISIS Fighter Shares Experiences On Tumblr Blog In English, Launches Pro-ISIS Telegram Channel In Swedish, July 22, 2016.

115   See MEMRI JTTM report Following Publication Of MEMRI Report About Pro-ISIS French Media Entity, Telegram Shuts Down Entity's Account – But It Is Relaunched Hours Later, July 21, 2016.

116   See MEMRI JTTM report Pro-ISIS Telegram Channel Distributes Photos Of U.S. Soldiers In Saudi Arabia Base, August 2, 2016.

117   See MEMRI CJL report Faces Of Death: On Twitter, Jihadis Distribute Photos Of 'Martyrs,' February 22, 2013; and Inquiry and Analysis No. 990, Faces Of Death – Part II: On Twitter, Jihadis Disseminate Death Photos Of Martyrs – Noting Their Beatific Smiles, Scent Of Musk Emanating From Their Bodies, And The Virgins Awaiting Them In Paradise; Other Popular Tweets Include Death Photo Of Boston Bomber Tamerlan Tsarnaev, Photo Of Basket Of Sweets Celebrating Martyr; Hashtags Created For Notable Martyrs, July 2, 2013.

118   See MEMRI CJL report WARNING – GRAPHIC: On Telegram, Jihadis Disseminate Death Photos Of Martyrs – Noting Their Beatific Smiles, Scent Of Musk Emanating From Their Bodies, And The Virgins Awaiting Them In Paradise, November 4, 2016.

119   See MEMRI CJL report WARNING – GRAPHIC: 'Strangers' Telegram Account Posts Pictures Of Brutal ISIS Executions For Western Sympathizers, September 8, 2016.

120   See MEMRI JTTM report In Wake Of Battle To Liberate Mosul, Pro-ISIS Writer Notes That 'Hundreds Of Mujahideen Cannot Stop' The Infidels, Condemns Online Jihadis For Staying Home, September 28, 2016.

121   MEMRI JTTM report ISIS Supporters Create Matrimonial Group On Telegram, July 6, 2016.

122   See MEMRI JTTM report ISIS Fighters, Supporters Following Launch Of Campaign To Retake Mosul: Claims That Enemy Being Pushed Back Due To Dozens Of Martyrdom Attacks; Calls For Lone Wolves To Attack New York, New Jersey, London, Paris, Ankara, Tehran As Revenge, October 18, 2016.

123   See MEMRI JTTM report Pro-ISIS Telegram Account Offers Tactical Tips To Lone Wolves In Saudi Arabia October 7, 2016.

124   See MEMRI JTTM report Pro-ISIS Telegram Channel Calls For Lone Wolves to Target Americans In Saudi Arabia, Saudi Intelligence And Security Personnel, October 25, 2016.

125   MEMRI JTTM report Self-Proclaimed American ISIS Fighter On Telegram Encourages Followers To Carry Out Lone-Wolf Attacks Instead Of Immigrating To Islamic State, June 1, 2016.

126   See MEMRI JTTM report ISIS Fighters, Including An American, A Frenchman And Russian, Praise Orlando Shooting, Call For More Attacks In West, U.S., Russia, June 20, 2016

127   MEMRI JTTM report Following Orlando Shooting, ISIS Supporters Gloat, Threaten Further Attacks On U.S. June 13, 2016.

128   See MEMRI JTTM report Following Terror Attack In Nice ISIS Supporters Post Banners Depicting Threats To France, U.S. Capitol, And Gloating Over France Tragedy, July 19, 2016.

129   See MEMRI JTTM report Video By Media Company Associated With ISIS Threatens Attacks On Western Cities – Including New York, Paris, Rome, Berlin, Moscow, July 14, 2016.

130   See MEMRI JTTM report French, Bosnian ISIS Fighters Urge Muslims In The U.S., Canada, Australia, Europe To Carry Out Attacks, June 27, 2016.

131   See MEMRI JTTM report Poem By Female Jihadi Praises Orlando Attack, Threatens U.S., July 7, 2016.

132   MEMRI JTTM report Pro-ISIS Libyan Media Activists Post Images Threatening Paris, Rome, May 26, 2016.

133   See MEMRI JTTM report Pro-ISIS Group Threatens Citizens Of Spanish-Speaking Countries: We Will Kill You Wherever You Are Unless You Stop Fighting Muslims, May 30, 2016.

134   See MEMRI JTTM report Hours Before The Euro 2016 Soccer Finals In Paris, ISIS French Media Share Threatening Posters Online, July 10, 2016.

135   See MEMRI JTTM report ISIS Operatives Distribute Kill List, Target List, Operational Tips For 'Lone Lions' In France And The West, August 15, 2016.

136   See MEMRI JTTM report Advice For 'Lone Wolves' On ISIS-Affiliated Telegram Channel: Poison The 'Infidels' Food, Air And Water, Spread Panic By Posting False Alarms, August 21, 2016.

137   See MEMRI JTTM report Pro-ISIS Group Urges "Lone Wolf" Operatives In Europe, Especially In France, To Carry Out Attacks Before They Are Arrested, September 4, 2016.

138   See MEMRI JTTM report Pro-ISIS Telegram Channel Calls on ISIS Supporters Everywhere To Use Cars, Knives, Rocks To Kill Infidels, October 31, 2016.

139   See MEMRI JTTM report Pro-ISIS Tech Channel On Telegram Posts Video Tutorial Showing How To Get Facebook Pages Suspended By Reporting Them As Belonging To Underage Children, October 20, 2016.

140   See MEMRI JTTM report Jihadi 'Help Desk,' Tech Channels On Telegram, Twitter Offer Tech Support, Tutorials, Up-To-Date Cyber Security Info, February 8, 2016.

141   See MEMRI CJL report Thousands Of ISIS Accounts On Telegram Warn: Don't Make Plans For Migration To Caliphate On Encrypted Messaging Platforms, July 22, 2016.

142   See CJL report American Jihadi Shares Safety And Encryption Tips On Telegram, May 4, 2016.

143   See MEMRI CJL report Pro-ISIS Telegram Channel Advertises For Volunteers To Translate, Write, Design, August 5, 2016.

144   See MEMRI JTTM report ISIS Releases Android App Teaching Supplications To Children, June 23, 2016.

145   See MEMRI JTTM report ISIS Engineers And Scientists Collaborate On Projects In Telegram Channel, March 18, 2016.

146   See MEMRI JTTM report ISIS Radio Launches Experimental Website On The Dark Web, July 5, 2016.

147   See MEMRI JTTM report Via Telegram, Jihadi Tech Group Offers Tutorial On Using Secure Messaging App Threema, March 9, 2016.

148   See MEMRI JTTM report To Avoid Suspension, ISIS Supporters Launch Seemingly Innocuous Telegram 'Sports News' Channel – For Promoting ISIS Channels, October 14, 2016.

149   MEMRI JTTM report ISIS Supporters Tighten Security Measures To Join ISIS Channels On Telegram, January 6, 2016.

150   See MEMRI JTTM report ISIS Supporters Move Conversation Offline, Create Chatrooms on Encrypted App 'Telegram', August 25, 2015.

151   See MEMRI JTTM report Pro-ISIS Telegram Channel Warns ISIS Supporters To Avoid Using WhatsApp, August 11, 2016.

152   See MEMRI JTTM report, Caliphate Cyber Army Promotes Latest Video Release On Telegram; Joins Forces With Pro-Palestinian Hackers AnonGhost, January 11, 2016.

153   See MEMRI CJL report Cyber Caliphate Army Back On Telegram, February 4, 2016.

154   See MEMRI CJL report New Cyber Caliphate Army Video Made With 'Cute Cut' For Apple iPad, Shows Group Editing Videos Using Same Software, March 14, 2016.

155   See MEMRI CJL report Caliphate Cyber Army Leaks New Jersey And New York Police Officer Data On Telegram, March 7, 2015.

156   See MEMRI CJL report On Telegram, Caliphate Cyber Army Calls To Celebrate Brussels Attacks Using Trending Hashtags, March 22, 2016.

157   See CJL report Caliphate Cyber Army Leaks Details On U.S. Police On Telegram, March 14, 2016.

158   See MEMRI JTTM report Pro-ISIS Hacking Group 'United Cyber Caliphate' Releases Personal Details Of U.S. State Department Employees April 25, 2016.

159   See MEMRI CJL report After Publishing State Dept Kill List, Caliphate Cyber Army Telegram Channel Shut Down; New Channel Launched Promising More Hacks, Mocking FBI, May 3, 2016.

160   See MEMRI CJL report Pro-ISIS Hacker Group Cyber Caliphate Army Back On Telegram, Announces New Collective, April 4, 2016.

161   See MEMRI JTTM report Pro-ISIS Hacking Group 'Caliphate Cyber Army' Threatens The U.S. In Upcoming Attack, See April 21, 2016.

162   MEMRI CJL report Warnings About Fake ISIS Apps 'Aimed At Infiltration' Circulated On Telegram, June 1, 2016.

163   See MEMRI CJL report On Telegram, Caliphate Cyber Army Posts Guide For Lone Wolf Attacks, June 24, 2016.

164   See MEMRI CJL report Pro-ISIS Hacking Group Posts 'Lone Wolf Scorecard' On Telegram, Marks Recent Stabbing, Bombing, Vehicular Attacks; Shooting, Poisoning, Beating Attacks Still Unmarked, July 26, 2016.

165   MEMRI JTTM Pro-ISIS Hacking Group Releases List Of 3,247 Individuals, Mostly Minority, Arrested In Dallas 2014-2016; Leaks List Of What It Claims Are Police Salaries, July 8, 2016.

166   See MEMRI JTTM report On Telegram, Pro-ISIS Hacking Groups Post 'Kill List' Featuring U.S., Australian, Canadian Citizens, June 7, 2016 .

167   See MEMRI CJL report On Telegram, Pro-ISIS Hacking Group Releases Kill List, Personal Contact Information Of USAF Air Mobility Command – AL, HI, NJ, PA, TX Officials Targeted,June 21, 2016 .

168   See MEMRI JTTM report Pro-ISIS Hacking Group Release Kill List With 1,693 Names Of 'Crusaders And Jews' In U.S., July 5, 2016.

169   See MEMRI JTTM report Pro-ISIS Hacking Group Caliphate Cyber Army Claims To Have Hacked Official Saudi Government Portal And Obtained Government Official Database, July 5, 2016.

170   See MEMRI CJL report Pro-ISIS Caliphate Cyber Army Links To New 'OPSEC IT' Telegram Channel, September 22, 2016.

171   See MEMRI JTTM report Pro-ISIS Hacking Group Releases List And Satellite Images Of U.S. Military Bases, June 7, 2016.

172   See MEMRI CJL report Caliphate Cyber Army Promotes Upcoming Operation On Telegram, January 5, 2016.

173   See MEMRI JTTM report Pro-ISIS United Cyber Caliphate Hacking Group Targets Australian Websites, April 14, 2016.

174   See MEMRI CJL report Pro-ISIS Hacker Group Cyber Caliphate Army Back On Telegram, Announces New Collective, April 4, 2016.

175   MEMRI CJL report In Response To Twitter's Crackdown On Extremist Accounts, Pro-ISIS Hacking Group Alleges It Hacked 5,000 Accounts On The Platform, August 23, 2016.

176   See MEMRI JTTM report Pro-ISIS Hacking Group Releases Kill List With Details Of Over 2,000 Officers And Soldiers On U.S. Military Bases, July 20, 2016.

177   See MEMRI CJL report Pro-ISIS Caliphate Cyber Army Distributes Request For Scientific Personnel On Telegram, August 31, 2016.

178   See MEMRI JTTM report Pro-ISIS Hacking Group 'Caliphate Cyber United' Leaks List Of Allegedly Prominent NYC Citizens, April 21, 2016.

179   See MEMRI JTTM report Pro-ISIS Hacking Group 'United Cyber Caliphate' Releases Personal Details Of U.S. State Department Employees April 25, 2016.

180   See MEMRI JTTM report Pro-ISIS Group Says It Hacked Database Of Saudi Defense And Interior Ministries, Releases Info Of Security Officials, April 25, 2016.

181   See MEMRI CJL report United Cyber Caliphate Hacks Vancouver Florist, May 11, 2016.

182   See MEMRI CJL report United Cyber Caliphate Posts On Telegram Announcement Of Hack Of 'Arkansas Library Database,' May 27, 2016.

183   See MEMRI CJL report United Cyber Caliphate Posts Social Media Safety Instructions, May 25, 2016.

184   SEE MEMRI JTTM report Pro-ISIS Hacking Group Releases 'Kill List' With 4,681 Names From Around The World – Including U.S., China, India, Australia, U.K, Canada - And From Microsoft, IBM, Walmart, Home Depot, Oracle,Yahoo, ExxonMobil, June 22, 2016.

185   See MEMRI JTTM report Pro-ISIS Hacking Group Releases Kill List Targeting Canadians, Containing Over 12,000 Entries, June 28, 2016.

186   See MEMRI JTTM report Pro-ISIS Hacking Group Releases Kill List Of 289 U.S. Army Corps Of Engineers Personnel, July 20, 2016.

187   See MEMRI JTTM report Pro-ISIS Hacking Group Releases Kill List Of Over 700 U.S. Army Personnel, July 25, 2016.

188   See MEMRI JTTM report Pro-ISIS Hacking Group Posts Infographic Highlighting Hacking Prowess, July 29, 2016.

189   See MEMRI JTTM report Pro-ISIS Hacking Group Release Kill List Featuring U.S. Air Force Personnel, August 3, 2016.

190   See MEMRI CJL report Pro-ISIS Hacker Group 'Sons Caliphate Army' On Telegram – Sharing Hacks, Threats, Jihad Content, March 31, 2016.

191   See MEMRI JTTM report Pro-ISIS Hacker Group 'Sons Caliphate Army' On Telegram Boasts Of Facebook, Twitter Hacks, Vows To Continue Cyber War, June 23, 2016.

192   MEMRI CJL report On Telegram, Pro-ISIS Hacking Group 'Sons Caliphate Army' Announces Twitter Accounts It Hacked Have Been Disabled, Mocks Facebook And Twitter CEOs Mark Zuckerberg And Jack Dorsey, September 12, 2016.

193   See MEMRI JTTM report In The Wake Of NY, NJ Bombings, Pro-ISIS Entity On Telegram Posts Tutorials On Building Pressure Cooker Bomb And Using Cellphones, Bluetooth For Remote Detonation, September 19, 2016.

194   See MEMRI CJL report Jihadi Hacking Group Cyber Kahilafah Uses Telegram to Inform Pro-ISIS Followers of Private Communication and Impeding Cyber Attack; Highlights Use of Online Information Sharing Platforms, April 4, 2016.

195   See MEMRI CJL report Qatar National Bank Database Hacked, Info Leaked Online Including Details On Al-Jazeera Reporters, Qatari Royal Family; Shared By Jihadi Hacker Group 'Cyber Khilafah', April 27, 2016.

196   See MEMRI CJL report Pro-ISIS Telegram Account Posts Dark Web Advertisements For Hiring Albanian Assassins And Hackers: 'You Don't Have To Pay 'Till The Job Is Done', June 20, 2016.

197   See MEMRI CJL report Cyber Kahilafah Telegram Channel Teases Hack Of London SCADA, August 5, 2016.

198   See MEMRI JTTM report Pro-ISIS Hacking Group Announces Plans To Publish Instructions For Remote Vehicle Operation Over Wi-Fi, September 26, 2016.

199   See MEMRI CJL report Pro-ISIS Cyber Kahilafah Hacking Group Previews Jehad Archives On The Darknet On Telegram, September 9, 2016.

200   See MEMRI CJL report Pro-ISIS Telegram Channel Forwards United Cyber Caliphate Instructions on VPN, April 28, 2016.

201   MEMRI CJL report Pro-ISIS Cyber Jihadi Group Kalachnikv E-Security Team Operates On Social Media, Publishes Addresses of Federal Reserve Board of Governors, April 4, 2016.

202   MEMRI JTTM report On Telegram, Pro-ISIS Hacking Group Posts Image Threatening Mark Zuckerberg, March 21, 2016.

203   MEMRI CJL report Kalachnikv E-Security Team Announces New Member From TheDarknet Nation, Provides Info On Various Security Software, May 4, 2016.

204   See MEMRI CJL report Kalachnikv E-Security Team Participates In #OpSaudiArabia, May 11, 2016.

205   See MEMRI CJL report Kalachinvk E-Security Team Hacks Indonesian Site, Threatens U.S. Government , May 25, 2016.

206   See MEMRI CJL report Caliphate Electronic Army Recruiting New Members, May 4, 2016.

207   See MEMRI CJL report Jihadi Hacking Group Cyber TeamRox (CTR) Active On Telegram, Facebook, March 9, 2016.

208   See MEMRI JTTM report Pro-ISIS Tech Channel Warns Of Alleged ISIS Telegram Groups Offering Hacking Tutorials, May 19, 2016.

209   See MEMRI CJL report Islamic Cyber Army Issues Security Directives On Telegram Following Shutdown Of Twitter Account, December 8, 2015.

210   See MEMRI JTTM report Designated Terror Group Ansar Al-Shari'a In Libya Begins Using Telegram To Disseminate Its News, May 6, 2015.

211   See MEMRI JTTM report Al-Qaeda Leader Ayman Al-Zawahiri Calls On Iraqi Muslims To Unite, Wage Guerilla Campaign To Rid Iraq Of Shi'ite-Crusader Occupation, August 25, 2016.

212   See MEMRI JTTM report Al-Qaeda Leader Ayman Al-Zawahiri Calls To Unite Jihadi Groups, Accuses ISIS Of Causing Schism And Harming Jihad, September 1, 2016.

213   See MEMRI JTTM report Al-Qaeda Gives Jabhat Al-Nusra Green Light To Leave Its Fold, July 28, 2016.

214   See MEMRI JTTM report GIMF Announces Its New Channel On 'Telegram', October 6, 2015.

215   See MEMRI JTTM report Al-Qaeda-Affiliated Media Company Global Islamic Media Front (GIMF) Creates Telegram Channel For Jihadi Content Related To Indian Subcontinent, August 9, 2016.

216   See MEMRI CJL report On Telegram, Global Islamic Media Front (GIMF) Shares Technical Advice For Aspiring Jihadis, February 24, 2016.

217   See MEMRI CJL report Al-Qaeda Media Wing Al-Sahab Active On Telegram, August 31, 2016.

218   See MEMRI CJL report Al-Qaeda Telegram Account Promotes Upcoming 9/11 Anniversary Video, September 6, 2016.

219   See MEMRI JTTM report AQAP Weekly Magazine Describes Hillary Clinton, Donald Trump As 'Full Of Flaws,' Says It Is Time For U.S. Prominence To Decline , October 26, 2016.

220   See MEMRI JTTM report Telegram Channel Disseminates Works Of Al-Awlaki, December 15, 2015

221   See MEMRI JTTM report Jihadis Promote New AQAP Channel On Secure Communications App 'Telegram', September 29, 2015.

222   See MEMRI CJL report Issue 20 Of AQAP Weekly 'Al-Masra' Announces Launch Of Official Telegram Channel, September 6, 2016.

223   See MEMRI JTTM report Telegram Channel Devoted To AQAP's 'Inspire' Magazine Posts Bomb-Making Instructions, March 18, 2016.

224   See MEMRI JTTM report In New Video, AQAP Presents Special Operations Brigade, Vows To Fight Until Liberation Of Al-Aqsa Mosque, July 13, 2016.

225   See MEMRI JTTM report Telegram Channel Distributes AQAP 'Inspire' Magazine, March 9, 2016.

226   See MEMRI JTTM report Pro-AQAP Telegram Channel Provides Instructions For Lone Wolf Attacks At Rio Summer Olympics; American, French, Israeli, British Athletes Singled Out, July 19, 2016.

227   See MEMRI CJL report In Wake Of Nice Attack, Telegram Channel Devoted To AQAP's Inspire Magazine Reposts Images Calling For Vehicular Attacks, July 15, 2016.

228   See MEMRI JTTM report Prominent Twitter Account Documenting 'U.S. Crimes In Yemen' Switches To Telegram, Expresses Frustration At Twitter's Repeated Suspensions, November 5, 2015.

229   See MEMRI JTTM report AQIM Announces On Telegram The Kidnapping Of Swiss Citizen In Mali, Issues Demands, January 27, 2016.

230   See MEMRI JTTM report AQIM Statement: Ivory Coast Attack Was Response To France's Military Campaign In Sahel – And Vengeance Against The Countries That Participate In French Military Operations In The Region, March 15, 2016.

231   See MEMRI JTTM report In New Audio Recording, Al-Shabab Leader Urges Somalis To Join Jihad, Calls Upon Muslims In Kenya, Ethiopia To Target 'Disbelievers' In Any Way Possible, July 13, 2016.

232   See MEMRI JTTM report Australian Jabhat Al-Nusra Cleric – And U.S.-Designated Global Terrorist – Active On Facebook, Twitter, Telegram, July 18, 2016.

233   See MEMRI JTTM report Senior Jabhat Al-Nusra Cleric, Australian Abu Sulayman, Praises The Syrians' Courage And Thanks Them For Their Hospitality, June 19, 2016.

234   See MEMRI JTTM Pro-Jabhat Fath-Al Sham Media Group Showcases Kurdish, Tajik, Syrian Fighters On Telegram, October 17, 2016.

235   See MEMRI CJL report Designated Terrorist Organization Hizbullah's Al-Manar TV Joins Telegram, October 11, 2016.

236   See MEMRI JTTM report English-Language Taliban Telegram Channels Publish News Updates, January 19, 2016.

237   See MEMRI CJL report English-Language Taliban Telegram Channel Offers Updates Via WhatsApp, January 21, 2016

238   See MEMRI CJL report Taliban-Affiliated Telegram Account Solicits Donations For Eid Al-Adha, September 9, 2016.

239   See MEMRI JTTM report On Telegram, Taliban Lashes Out At Brussels Conference On Afghanistan , October 6, 2016.

240   See MEMRI JTTM report On Telegram, Taliban Issues Statement, Threats Of U.S. Invasion Of Afghanistan, October 11, 2016.

241   See MEMRI JTTM report Turkestan Islamic Party (TIP) Launches Telegram Channel, January 13, 2016.

242   See MEMRI JTTM report Turkestan Islamic Party (TIP) In Syria Urges Xinjiang Muslims To Join The Jihad In New Video, July 12, 2016.

243   See MEMRI CJL report Turkestan Islamic Party (TIP) Launches New Channel On Telegram, August 10, 2016.

244   See MEMRI JTTM report Turkestan Islamic Party (TIP) Releases New Music Video In Uyghur: 'Rise Up, Oh Turkestan,' September 6, 2016.

245   See MEMRI JTTM report ISIS Releases New Song In Uyghur, Titled 'Religion Of Ibrahim' August 24, 2016.

246   See MEMRI CJL report On Its Telegram Account, Hamas's Al-Qassam Martyrs Brigade Stresses Jihad And Martyrdom, March 29, 2016.

247   See MEMRI JTTM report Telegram Account Reposts Document Offering Stabbing Tactics, Bomb Making Instructions For Palestinians Planning On Attacking Jews, January 26, 2016.

248   See MEMRI JTTM report On Twitter And Telegram – Campaign To Fund Weapons, Fighters To Target Jews, November 13, 2015.

249   See MEMRI JTTM report Fundraising Campaign To Arm Jihadis In Gaza Solicits Donations Via Bitcoin, July 8, 2016.

250   See MEMRI CJL report Al-Qaeda, Jihadis Infest the San Francisco, California-Based 'Internet Archive' Library, August 20, 2011.

251   Archive.org/about.

252   See MEMRI CJL report Jihadi Telegram Channels Using Internet Archive To Spread Content From ISIS, Al-Qaeda, And Other Groups, November 3, 2016.

253   See MEMRI CJL report Telegram 'Nasheed Gallery' Channel Shares Songs Featuring Voices Of Jihadis, 'Martyrs', July 11, 2016.

254   See MEMRI JTTM report Unknown Egyptian Militant Group, Jama'at Al-Murabitoon, Uses Secure Messaging App 'Telegram', October 5, 2016.

www.ingramcontent.com/pod-product-compliance
Lightning Source LLC
Chambersburg PA
CBHW041443210326
41599CB00004B/119